Optimizing the Contributions of Air Force Civilian STEM Workforce

SHIRLEY M. ROSS, REBECCA HERMAN, IRINA A. CHINDEA,
SAMANTHA E. DINICOLA, AMY GRACE DONOHUE

Prepared for the United States Air Force
Approved for public release; distribution unlimited

PROJECT AIR FORCE

For more information on this publication, visit www.rand.org/t/RR4234

Library of Congress Cataloging-in-Publication Data is available for this publication.

ISBN: 978-1-9774-0442-8

Published by the RAND Corporation, Santa Monica, Calif.

© Copyright 2020 RAND Corporation

RAND® is a registered trademark.

Support RAND

Make a tax-deductible charitable contribution at

www.rand.org/giving/contribute

www.rand.org

Preface

The U.S. Air Force's ability to accomplish national security goals relies heavily on research advances in the science, technology, engineering, and mathematics (STEM) fields. The current shortage of STEM professionals has a direct impact on how the Air Force is able to carry out its mission. Addressing the gap in the Air Force's civilian STEM workforce and optimizing the productivity of its existing civilian STEM employees falls squarely within the Air Force's responsibility. Because of concerns over the shortage of civilian STEM professionals, especially those with advanced degrees, Air Force leadership asked RAND Project AIR FORCE (PAF) to explore the existing academic and professional literature to gain insights into how organizations such as the Air Force should manage, support, and organize their current civilian STEM workers to best leverage their talents and thereby maximize performance.

PAF engaged in an extensive survey of the relevant literature to answer the above question. First, we provided a brief overview of the differences between modern knowledge organizations, in contrast to traditional manufacturing or industrial organizations. Second, we described the characteristics of work that most appeal to STEM workers and drive their productivity. Third, we discussed human-capital functions that relate to the performance of STEM workers. Fourth, we discussed the changes in organizational structure most likely to foster STEM employees' productivity and innovation. Finally, the last section of this report summarizes our findings and recommendations.

The research reported here was sponsored by the U.S. Air Force Manpower, Personnel and Services (AF/A1) and conducted within the Manpower, Personnel & Training Program of RAND Project AIR FORCE as part of a fiscal year 2018 project "Continuing U.S. Air Force Human Capital Strategic Initiatives."

RAND Project AIR FORCE

RAND Project AIR FORCE (PAF), a division of the RAND Corporation, is the U.S. Air Force's federally funded research and development center for studies and analyses. PAF provides the Air Force with independent analyses of policy alternatives affecting the development, employment, combat readiness, and support of current and future air, space, and cyber forces. Research is conducted in four programs: Force Modernization and Employment; Manpower, Personnel, and Training; Resource Management; and Strategy and Doctrine. The research reported here was prepared under contract FA7014-16-D-1000.

Additional information about PAF is available on our website: www.rand.org/paf/

This report documents work originally shared with the U.S. Air Force on December 10, 2018. The draft report, issued on November 28, 2018, was reviewed by formal peer reviewers and U.S. Air Force subject-matter experts.

Contents

Preface iii

Summary vi

Acknowledgments xi

Abbreviations xii

1. Introduction 1

    Background: The Knowledge-Based Economy and the Organization of Work 3

    The STEM Workforce and National Security 5

    Report Objectives 7

    Organization of This Report 7

2. Optimizing the Alignment Between Work and STEM Professional Characteristics 8

    Organizational Culture and Climate 9

    Autonomy 11

    Collaboration and Work Design 13

    Focus on Substantive Work 16

    Flexible Work Arrangements 17

    Women in STEM Fields 20

3. Human Capital Functions 24

    Development 24

    Rewards and Recognition 25

    Career Advancement 27

    Performance Management 28

4. The Role of Organizational Structure in Optimizing Performance of STEM Workers 30

    The Structure of Knowledge-Based Organizations 30

    Innovation Cells 31

    Hyperspecialization 33

5. Conclusions and Recommendations 35

    Aligning Work and STEM Professionals' Characteristics 35

    Human Capital Functions 36

    Organizational Structure Optimizing the Performance of STEM Professionals 36

    Challenges to Implementation 37

Selected Bibliography 38

Summary

As the United States continues its entrenchment into a knowledge-driven economy, the quantity and quality of professionals with undergraduate and graduate degrees in the fields of science, technology, engineering, and math (STEM) have continued to be the focus of leaders in the public and private sectors. While the demand for qualified STEM professionals has increased continuously in the past decade, the share of U.S. students earning STEM undergraduate and graduate degrees has declined, translating into a shortage of STEM professionals with the desirable qualifications.[1]

Given the scarcity and importance of STEM professionals, it is especially important that the U.S. Air Force, as well as the U.S. Department of Defense (DoD), maximize the impact of the existing civilian STEM workforce. This report explores the question of how organizations such as the Air Force should manage, support, and organize civilian STEM professionals to best leverage their individual and collective talents and thereby maximize performance, productivity, and innovation.

Consequently, this report aims to examine and summarize findings from the scholarly and professional literature related to optimizing the effectiveness and productivity of professionals engaged in STEM occupations. We are particularly interested in approaches that could maximize organizational outcomes of STEM workforces in national security organizations, such as the Air Force in particular and DoD and its components in general. Although this report focuses mainly on the civilian STEM workforce, some of the findings and recommendations are likely to be applicable to noncivilian STEM professionals across the Air Force and DoD. In light of estimated future retirements and overall concerns with the federal government's ability to recruit talent in a timely fashion, the primary focus of this report is on improving performance outcomes of the current civilian STEM employees within military organizations. Hence, in this study, we do not present an in-depth discussion of recruiting and retention of STEM employees—which is well covered elsewhere in the literature—even though many of the suggestions that we propose would be prime factors for the retention of STEM workers alongside the improvement in their productivity.

Managing an organization's STEM workforce and optimizing its productivity is considered one of the greatest challenges for organizational leadership. In general, organization design is still shaped largely by best practices for managing clerical work, with many organizations struggling to support and manage STEM workers, who have unique motivations and needs. In the specific case of the Air Force, administrative structures are in place that need further

---

[1] In this report, we use interchangeably the terms *STEM workforce*, *STEM professionals*, and *STEM workers*, who are individuals who hold at least a bachelor's degree in one of the STEM fields and are employed in a STEM job.

improvement to optimize STEM workers' motivation and productivity, with the organization's structure, culture, and core values all greatly affecting STEM workers' motivation, productivity, and innovation.

Background

Since the late 1980s, economies throughout the world have started to transition away from the manufacturing business model that prevailed during the 19th and most of the 20th centuries to a knowledge-based business model. Two factors drove the transformation: advanced information technology and the need for industry to innovate and become more entrepreneurial. Management consultant and academic Peter Drucker—whose work laid out the foundations of management science in the modern corporation—defined the information-based organization as one "composed largely of specialists, who direct and discipline their own performance through organized feedback from colleagues, customers, and headquarters."[2]

Optimizing the Alignment Between Work and STEM Professional Characteristics

Prior research has identified four main characteristics that should be considered when developing and maintaining a STEM workforce: autonomy of STEM employees in selecting and managing their work, collaboration with specialists having complementary knowledge, focus on substantive work rather than management or administrative tasks, and flexible work arrangements (FWAs). Knowing the characteristics, needs, and expectations of a STEM professional helps organizations rethink and redesign the work setting, organizational culture, and climate that would maximize their efforts and foster innovation.

- *Autonomy* comprises two components: authority to (1) select the focus of one's work and (2) manage one's own work processes. Autonomy appears especially important to STEM workers and relates to higher performance. In fact, offering more autonomy in selecting work may motivate STEM professionals, while reducing autonomy may demotivate them. Hence, in hierarchical organizations such as the Air Force and DoD, where STEM workers are less likely to be able to select the focus of their work, allowing them to have a strong input into projects and autonomy in managing how they complete the work is likely to compensate for the lack of ability to select work focus in most situations.
- *Collaboration* is highly valued by many STEM workers, who rely heavily on learning from peers in the workplace and experts from other organizations. They view personal relationships as an important way to transfer knowledge and consider on-the-job problem-solving and colleague interaction as the two most important professional growth activities. Repeated formal and informal interactions among STEM researchers contribute

---

[2] Peter F. Drucker, "The Coming of the New Organization," *Harvard Business Review*, January 1988.

to building mutual trust, resulting in interpersonal exchanges of resources across organizational boundaries, which are critical to fostering innovation.

- *Focus on substantive work* may highly motivate STEM workers, who desire to have work they view as challenging and interesting and who would rather avoid repetitive, routine work that is often associated with administrative tasks. Next to autonomy in choice of work, the elimination of routine tasks is of critical importance to optimizing the STEM workforce.

- *Flexible work arrangements* (FWA) allow employees varying levels of control over the time during which and the place where work occurs. As STEM work is often project-based—requiring coordination with team members and access to specific resources or technologies—FWA for STEM work might be difficult to implement. Furthermore, in defense organizations such as the Air Force, in which work involves accessing and handling classified information, FWA arrangements are more difficult to implement. To balance FWA with the requirements of lab or office presence, the Air Force should consider setting up a schedule of research activities. Under such a schedule, STEM workers would be able to coordinate their on-site presence, while still taking part of the time advantage of FWA.

Autonomy, collaboration, focus on substantive work, and FWAs are also important for DoD women in STEM. On the one hand, women are underrepresented in such occupations as engineering, computer science, and the physical sciences, while, on the other, women in DoD are present in lower numbers in the civilian workforce of each military service. This overlapping underrepresentation of women warrants a deeper look at the ways in which the productivity of DoD women in STEM can be optimized. Alongside supporting the four generic characteristics of the STEM workforce, DoD STEM organizational culture and climate also need to incorporate and consider factors that are women-specific and increase the productivity of women in STEM occupations: changes in stereotypes associated with the STEM work environment and with women's abilities, presence of women role models, opportunities for professional growth, a sexual harassment–free work environment, and family-friendly policies.

Human Capital Functions

Human capital functions are often portrayed by a life-cycle model that consists of seven major phases: workforce planning, talent acquisition, workforce development, performance management, rewards and recognition, career planning, and succession planning. Of these seven phases, four tap into the individual workers' needs and contributions: development, rewards and recognition, performance management, and career planning. They can be powerful levers when tailored to the needs of STEM workers:

- *Development*: Professional growth is a powerful motivator for STEM workers. By providing opportunities for professional growth, an organization can create a more positive work environment and increase workers' self-motivation.

- *Rewards and recognition*: Both extrinsic and intrinsic reward and recognition systems can significantly boost STEM workers' productivity. When properly implemented, rewards and recognition help strengthen STEM employees' efficiency and motivation.
- *Career advancement*: To retain the technically oriented high performers in the areas of the organization where they can contribute most, opportunities for career advancement and promotion should be equally available for both technical and managerial tracks, with the reward system for the former not lagging behind the latter in terms of status and financial compensation.
- *Performance management*: An effective performance-management system for STEM employees considers the individual worker's inclination toward pursuing a lifelong technically focused career or interest in transitioning into management roles. The incentives mechanism would ideally be designed to reward the areas in which individual STEM employees excel (technical or administrative).

Role of Organizational Structure in Optimizing Performance of STEM Workers

U.S. organizations are overwhelmingly organized around traditional hierarchies. However, a more decentralized and flat structure that connects autonomous task forces or units (e.g., innovation cells) in a networklike fashion is more likely to increase the productivity of its STEM workforce. A networked structure within and across innovation cells is likely to not only stimulate innovation but also increase productivity by facilitating communication and collaboration.

- *Innovation cells* are stand-alone units that are structured differently, operate differently, and have different expectations for outcomes than the parent organizations. They are created to best leverage the workforce (in this case, the STEM workforce), increase productivity, and encourage innovation among its ranks.
- *Hyperspecialization* of STEM workers benefits the organization in terms of quality, speed, and cost. However, hyperspecialization requires additional activities to break down larger tasks into discrete subtasks and may entail some risks born from the lack of consistent regulations to govern the work across topics and countries. There are also concerns that hyperspecialization might stifle innovation.

We conclude this report by recommending that the Air Force establish, in a limited way, a separate, simplified, or even flat organizational structure that facilitates collaboration and knowledge sharing across the STEM workforce. Setting up autonomous cells or task forces that interact with one another across networks rather than in hierarchies is likely to provide the STEM workforce with greater autonomy. By promoting an organizational culture and climate that take into account the particular needs of STEM work, such as autonomy, collaboration, focus on substantive work, and FWAs, the Air Force is more likely to promote creativity, innovation, and productivity across its STEM workforce. In addition, to fully benefit from the skills and capability that women STEM workers can contribute, the Air Force should consider increasing the number of successful women who could serve as role models, providing opportunities for professional growth and family-friendly policies, and ensuring a stereotype- and sexual harassment–free work environment.

Lastly, in terms of human capital functions, the Air Force might consider expanding the professional-development opportunities offered to civilian STEM employees, including STEM programs currently reserved for uniformed service members. Furthermore, the Air Force might consider bringing the compensation of its STEM workforce in line as much as possible with private-sector compensation, while allowing for autonomy and flexibility, as well as for performance-management and career-advancement paths that take into account individual interests in promotion and in the pursuit of different career tracks. While there is evidence that these aspects are likely to improve the productivity of STEM workers in general, we recommend that the Air Force conduct its own independent study to determine which factors and in what combination are likely to have the highest impact on the productivity of civilian STEM workers in the service.

Acknowledgments

We are grateful to the many people who were involved in this research. Specifically, we thank our Air Force sponsor, Daniel R. Sitterly, Principal Deputy Assistant Secretary for Manpower and Reserve Affairs, for initiating this effort, and Gwendolyn R. DeFilippi, Assistant Deputy Chief of Staff for Manpower, Personnel and Services, for her support in completing the process.

We thank Ray Conley, director of RAND Project AIR FORCE's Manpower, Personnel and Training Program, and Kirsten Keller, associate director of the Manpower, Personnel and Training Program, who provided steady guidance and unwavering support throughout the research and review process.

Chaitra Hardison and Bob Rogers kindly agreed to serve as internal and, respectively, external reviewers for this report. Their comments and recommendations greatly improved the quality of our analysis.

Last but not least, we thank our RAND colleagues Elizabeth Hammes for her assistance with the research and collection of many of the articles we used in this report, and Linda Theung for her input and infinite patience during the editing process.

Abbreviations

DoD	U.S. Department of Defense
FTF	face-to-face
FWA	flexible work arrangement
R&D	research and development
STEM	science, technology, engineering, and math
USAF	U.S. Air Force

1. Introduction

As the United States continues its entrenchment into a knowledge-driven economy, the quantity and quality of professionals with undergraduate and graduate degrees in the fields of science, technology, engineering, and math (STEM) have continued to be the focus of leaders in the public and private sectors. The U.S. Bureau of Labor Statistics predicted an 8.8-percent increase in STEM jobs between 2018 and 2028, a growth rate higher than the predicted 5 percent growth rate for jobs in fields outside STEM,[3] while historically, from 1990 to 2016, overall employment in STEM occupations rose 79 percent, from 9.7 million to 17.3 million.[4] Meanwhile, the share of U.S. students earning STEM undergraduate and graduate degrees in the last 25 years has declined and continues to stagnate,[5] translating into a shortage of qualified STEM professionals.

Given the scarcity and importance of STEM professionals, it is especially important that the U.S. Air Force, as well as the U.S. Department of Defense (DoD), maximize the impact of the existing STEM workforce. This report explores the question of how organizations such as the Air Force should manage, support, and organize civilian STEM professionals to best leverage their individual and collective talents and thereby maximize performance, productivity, and innovation.

DoD currently does not have an official definition for the STEM workforce.[6] Studies of the STEM workforce conducted by various government agencies have varying degrees of agreement about which occupations are included in STEM, the minimum educational requirements for STEM professions,[7] and the statistics used to generate the estimated size of the U.S. STEM

---

[3] U.S. Department of Labor, U.S. Bureau of Labor Statistics, "Employment Projections: Employment in STEM Occupations 2018–2028," last updated September 4, 2019.

[4] Pew Research Center analysis of U.S. Census Bureau data from 1990 to 2016, presented in Lisa McBride, "Changing the Culture for Women and Underrepresented Groups in STEM+M," *Insights into Diversity*, August 22, 2018.

[5] U.S. Congress Joint Economic Committee, *STEM Education: Preparing for the Jobs of the Future*, a report by the Joint Economic Committee Chairman's Staff Senator Bob Casey, Chairman, Washington, D.C., April 2012.

[6] National Academy of Engineering and National Research Council, *Assuring the U.S. Department of Defense a Strong Science, Technology, Engineering, and Mathematics (STEM) Workforce*, Washington, D.C.: National Academies Press, 2012, p. 37; for a discussion of Air Force definition of STEM, see Lisa M. Harrington, Lindsay Daugherty, S. Craig Moore, and Tara L. Terry, *Air Force-Wide Needs for Science, Technology, Engineering, and Mathematics (STEM) Academic Degrees*, Santa Monica, Calif.: RAND Corporation, RR-659-AF, 2014.

[7] For instance, Laurence Shatkin defines STEM occupations "as those requiring knowledge of or skill with science, technology, engineering, or math with at least two-years of postsecondary study or training" (Shatkin cited in Rich Feller, "Advancing the STEM Workforce Through STEM-Centric Career Development," *Technology and Engineering Teacher*, Vol. 71, No. 1, September 2011, p. 10).

workforce.[8] For the purposes of this report, *STEM* broadly refers to technical jobs in the fields of science, technology, engineering, and mathematics. Professions include engineers, mathematicians, computer scientists and cybersecurity specialists, data scientists, and life and physical scientists.[9] We define the *STEM workforce* as made up of individuals who hold at least a bachelor's degree in one of the STEM fields and are employed in a STEM job. Although the characteristics of the STEM workforce presented in this report are likely to be equally applicable to those STEM workers who hold only a STEM college degree or less,[10] we hope our findings will assist the Air Force's and DoD's efforts to stimulate the productivity of STEM workers who hold a STEM graduate degree for the following two reasons:

- On average, about 30 percent of DoD's civilian STEM workforce have graduate degrees (of which 5 percent are doctoral degrees).[11] DoD and the Air Force are experiencing difficulties in hiring at doctoral level, and there is a supply shortage in extremely specialized areas that need to be filled by STEM graduates with advanced degrees.[12]
- The rising profile of advanced technologies such as artificial intelligence and autonomous systems in the context of an increasingly competitive security environment raises the importance of both quantity and quality in STEM hiring, with graduate degree holders sharpening DoD's competitive edge in a highly specialized technological environment.

We consider that STEM workers represent a subset in the wider category of *knowledge workers*, who—according to management consultant and academic Peter Drucker—are "specialists who direct and discipline their own performance through organized feedback from colleagues, customers, and headquarters."[13] With the emergence in the 1980s of a new economic-development paradigm—knowledge economy—the emphasis shifted to "the role of knowledge creation and distribution as the primary driver in the process of economic growth."[14] Knowledge workers represent the main value creators in the knowledge economy. Together with

---

[8] For a more detailed discussion, see National Academy of Engineering and National Research Council, 2012, p. 37, and Yi Xue and Richard C. Larson, "STEM Crisis or STEM Surplus: Yes and Yes," *Monthly Labor Review*, Vol. 138, May 2015.

[9] Dennis Vilorio, "STEM 101: Intro to Tomorrow's Jobs," *Occupational Outlook Quarterly*, Spring 2014, pp. 2–12.

[10] According to data provided by the Defense Manpower Data Center, between 2001 and 2011, approximately 23 percent of the DoD STEM workforce had less than a bachelor's degree, primarily reflecting a high percentage of computer scientists and mathematical scientists with no bachelor's degree. For details, see National Academy of Engineering and National Research Council, 2012, pp. 52–54.

[11] National Academy of Engineering and National Research Council, 2012, p. 54.

[12] National Academy of Engineering and National Research Council, 2012; Harrington et al., 2014; Timothy Coffey, "Building the S&E Workforce for 2040: Challenges Facing the Department of Defense," Washington, D.C.: Center for Technology and National Security Policy, National Defense University, July 2008; Xue and Larson, 2015.

[13] Peter F. Drucker, "The Coming of the New Organization," *Harvard Business Review*, January 1988, p. 45.

[14] Richard G. Harris, "The Knowledge-Based Economy: Intellectual Origins and New Economic Perspectives," *International Journal of Management Reviews*, Vol. 3, No. 1, March 2001, p. 21.

other white-collar workers and professionals—such as lawyers and academics—STEM workers represent one of the major components of the knowledge workforce.

Because STEM workers represent a subcategory of knowledge workers, many of the findings from the literature on the characteristics of the knowledge workforce are generalizable to this narrower subcategory that is the STEM workforce.[15] Using a similar logic and given the prevailing inconsistencies in the literature about the level of education of those in the STEM workforce,[16] we consider that the findings of broader studies that include STEM bachelor's degree holders or less are likely to be generalizable to the narrower category of STEM workers with postgraduate degrees. Therefore, unless specified otherwise, we use the terms *STEM workforce*, *STEM professionals*, and *STEM workers* interchangeably to refer to individuals who hold at least a bachelor's degree in one of the STEM fields and are employed in a STEM job.

Background: The Knowledge-Based Economy and the Organization of Work

Since the late 1980s, economies throughout the world in general and in the United States specifically have started to transition away from the manufacturing business model that prevailed during the 19th and most of the 20th centuries to a knowledge-based business model. According to Drucker, two of the factors driving the transformation were advanced information technology and the need for industry to innovate and become more entrepreneurial.[17] Similarly, for military organizations, these factors, together with the need to successfully face the challenge of near-peer competitors such as China and Russia,[18] had an impact on the Air Force's need to better integrate knowledge workers in general and STEM workers specifically and incorporate knowledge-based practices within the organization.

The challenge of the task is not insignificant. The Air Force—a military organization—is an archetypal command-and-control entity, with a decades-long mission and organizational culture closely tied to a hierarchical command structure. The military command-and-control organization has actually inspired, in the last century, the organization of businesses and the second evolution of business organizations.[19] As Drucker observed, the move to a knowledge-based organization represented the driver behind a third evolutionary phase in the structure of organizations. However, military organizations such as the Air Force have largely remained

---

[15] Furthermore, many of the sources used in this report use interchangeably the term *knowledge workers* when referring to engineers, R&D scientists, and others in technical occupations.

[16] Some studies focus exclusively on STEM graduates while others also include bachelor's degree holders and nondegree holders as part of the STEM workforce.

[17] Drucker, 1988, pp. 45–46.

[18] The White House, *The National Security Strategy of the United States of America*, Washington, D.C., December 2017, p. 2.

[19] Drucker, 1988, p. 45.

organized along highly hierarchical and bureaucratic lines, often lacking the agility of flat, highly reactive, and innovative tech organizations. Hierarchical organizations' lack of agility makes it difficult for most of them to compete successfully in the knowledge-based and rapid-innovation culture of the 21st century.[20]

According to Drucker, the first evolution took place between 1895 and 1905 in the context of the industrial revolution, as ownership and management came to be seen as distinct, and management was recognized as work in its own right. The second evolution occurred about two decades later with the appearance of the modern command-and-control organization—largely inspired by military organizations[21]—and its delineations between policy and operations and with the addition of such professional functions as personnel management and budget and finance. With the rapid advancement in information and communication technologies, the third evolution manifested in the move away from this command-and-control structure toward the organization of knowledge specialists.[22] In turn, this would create challenges when it came to developing rewards, recognition and career opportunities for knowledge workers, and devising a management structure for an organization of task forces.[23]

From the late 1980s until the present, organizations—including military ones such as the Air Force and DoD—have continued to struggle with realigning their traditional practices and processes to a workforce largely made up of knowledge workers, of which STEM professionals are a subset. Catchphrases linked to efforts to organize STEM professionals are widespread in the media: dual-career tracks, innovation cells, innovation corps. Some of these terms denote approaches rooted in human capital practices, such as *dual career track*, which is a career ladder for scientists and engineers that tracks not to general management but to the most senior scientific and technical roles in the organization. Other terms, such as *innovation cells* or *innovation corps*, denote approaches in how the organization itself is structured to best leverage selected professionals.

In the past three decades, a rich body of scholarly and professional literature has explored the challenges and solutions associated with integrating and leveraging the talents of knowledge workers broadly and of STEM workers specifically. For military organizations, which still find themselves in Drucker's second evolutionary phase of organizational structure, the challenge of optimizing the productivity of their STEM workforce is even more daunting, as it is highly pressing in the existing international security environment.

---

[20] Dongil D. Keum and Kelly E. See, "The Influence of Hierarchy on Idea Generation and Selection in the Innovation Process," *Organization Science*, Vol. 28, No. 4, July 2017, pp. 653–669; Kate Crawford, Helen M. Hasan, Leoni Warne, and Henry Linger, "From Traditional Knowledge Management in Hierarchical Organizations to a Network Centric Paradigm for a Changing World," *Emergence: Complexity and Organization*, Vol. 11, No. 1, 2009, pp. 1–18; Tim Kastelle, "Hierarchy Is Overrated," *Harvard Business Review*, November 20, 2013.

[21] Drucker, 1988, p. 45.

[22] Drucker, 1988, p. 53.

[23] Drucker, 1988, p. 50.

The STEM Workforce and National Security

U.S. national security relies heavily on STEM research advances. The U.S. military's effectiveness in future conflicts, its ability to protect its citizens, and the U.S. government's broader capacity to carry out basic missions such as humanitarian efforts and science-based activities, all depend heavily on continued advances in the U.S. technology base.[24]

Both DoD and the Air Force have directly experienced a supply shortage of STEM professionals, most significantly in niche, extremely specialized areas in which STEM graduates with advanced degrees are most needed.[25] Xue and Larson revealed that hiring STEM workers with bachelor's degrees is "relatively easy," but shortages persist at the master's and doctorate level.[26] DoD has also reported a shortage of STEM workers in certain specialty fields, including cybersecurity and intelligence.[27]

These trends have been confirmed in the annual Industrial Capabilities report issued in March 2018 by DoD's Office of Manufacturing and Industrial Base Policy, which highlighted that aerospace and defense companies "are being faced with a shortage of qualified workers to meet current demands as well as needing to integrate a younger workforce with the 'right skills, aptitude, experience, and interest to step into the jobs vacated by senior-level engineers and skilled technicians' as they exit the workforce."[28]

Gaps in DoD's STEM workforce exist for several reasons. One partial explanation is that defense industry positions involve strict citizenship and security clearance requirements.[29] A second and related explanation is the decline in the share of U.S. citizens earning advanced STEM degrees. For instance, in 2009, U.S. citizens earned only 54 percent of the STEM doctorates awarded in the United States, compared with earning 74 percent of the doctorates awarded in 1985.[30] However, during this time, both the percentage of foreign nationals earning STEM degrees in the United States and the demand for qualified STEM workers have continued to increase. Another explanation is found in the aging of the current DoD STEM workforce. A

---

[24] National Academy of Engineering and National Research Council, 2012; The White House, *National Security Strategy of the United States (2010)*, Washington, D.C., May 2010.

[25] National Academy of Engineering and National Research Council, 2012; Coffey, 2008; Xue and Larson, 2015.

[26] Xue and Larson, 2015.

[27] National Academy of Engineering and National Research Council, 2012.

[28] Office of the Under Secretary of Defense for Acquisition and Sustainment and Office of the Deputy Assistant Secretary of Defense for Manufacturing and Industrial Base Policy, *Report to Congress: Fiscal Year 2017 Annual Industrial Capabilities Report to Congress*, Washington, D.C., March 2018, p. 8. The Emerson *Fourth Annual Survey* also supports these findings; see Emerson, "Emerson Survey: 2 in 5 Americans Believe the STEM Worker Shortage is at Crisis Level," August 21, 2018.

[29] National Academy of Engineering and National Research Council, 2012; Xue and Larson, 2015.

[30] U.S. Congress Joint Economic Committee, 2012.

growing number of STEM employees with advanced degrees are reaching retirement eligibility.[31]

Most significantly, the STEM research field—which predominantly includes STEM workers with advanced degrees—is diversifying, with a growing number of companies outside DoD and its contracting community drawing away advanced-degree STEM workers.[32] Moreover, DoD's STEM research budget—a subset of DoD and Air Force budgeted activities—is considerably smaller than it once was, and it can no longer significantly influence the size and skills of the STEM workforce through large-scale hiring.[33]

Recruiting STEM professionals into DoD and retaining them is a clear and documented challenge, and there is substantial effort to improve DoD STEM recruitment and retention. However, this is not the only challenge that DoD faces to ensure access to cutting-edge STEM capabilities. Managing the STEM workforce for maximum effectiveness may be an equally important challenge to securing STEM talent. Creating the conditions that support effective use of STEM professionals is a potentially powerful lever that has been seriously underresearched.

In light of these conditions, the primary focus of this report is on improving performance outcomes, such as increased rate of innovation and collective productivity, of existing civilian STEM employees within organizations such as DoD and the Air Force. Therefore, this report does not present an in-depth discussion of recruiting and retention of STEM employees, even though many of the suggestions we propose would be prime factors to retain STEM workers, as well as to improve their productivity. We also acknowledge that while the focus of the report is on the civilian STEM workforce, some of the findings are likely to be applicable to the active-duty STEM personnel.

The importance of improving the management of the existing STEM workforce was outlined in the 2012 study of the Committee on Science, Technology, Engineering, and Mathematics Workforce Needs for DoD and the U.S. Defense Industrial Base, which concluded that

> [t]he fundamental issue is quality, agility, and skills mix in the DoD STEM workforce. . . . Less-than-effective management of the DoD's STEM workforce inhibits recruiting and retention by limiting career growth, underutilizing employee skills, and constraining the available pool of talent.[34]

---

[31] National Academy of Engineering and National Research Council, 2012.

[32] National Academy of Engineering and National Research Council, 2012; Xue and Larson, 2015; U.S. Congress Joint Economic Committee, 2012.

[33] National Academy of Engineering and National Research Council, 2012. It is worth noting, however, that under President Trump's fiscal year 2019 budget request, the level of spending on DoD's Research, Development, Test, and Evaluation (RDT&E) accounts was to be increased to $90 billion and return "to the post–Cold War peak last seen almost a decade ago." See Will Thomas, "FY19 Budget Request: Defense S&T Stable as DOD Focuses on Technology Transition," *American Institute of Physics*, No. 20, February 23, 2018.

[34] National Academy of Engineering and National Research Council, 2012, p. 115.

Managing an organization's STEM workforce and optimizing its productivity are considered some of the greatest challenges for organization leadership.[35] Organizations struggle to support and manage STEM workers, who have unique motivations and needs born out of the focused intellectual and innovative processes in which they engage as part of their professional obligations. In typical command-and-control organizations such as the Air Force and DoD, the adoption of structural changes such as those currently being used in progressive high-tech firms are likely to optimize the motivation and productivity of STEM workers,[36] who belong to the most recent wave of the knowledge-workforce. The organization's structure and culture greatly affect the optimization of STEM workers' motivation, productivity, and innovation.[37]

Report Objectives

The purpose of this report is to examine and summarize findings from the scholarly and professional literature related to optimizing the effectiveness and productivity of professionals engaged in STEM occupations and careers. We are particularly interested in approaches that could maximize organizational outcomes of civilian STEM workforces in national security organizations, such as the Air Force. Furthermore, we argue that the findings of our report have wider implications beyond the Air Force and the civilian side and are more broadly applicable to the STEM workforce across DoD.

Organization of This Report

In this report, we examine the optimization of the STEM workforce from three perspectives. First, in Chapter Two, we present findings related to characteristics of the knowledge work itself, looking at how work can be best aligned with the needs and motivations common to STEM professionals. Chapter Three concerns human capital functions such as development, compensation, career planning, and performance management and how they can optimize STEM worker productivity. Chapter Four focuses on the structure of the organization itself, including stand-alone entities such as innovation cells, which have recently become a popular structure to stimulate innovation. The report concludes with recommendations and a summary of findings.

---

[35] Drucker, 1988; Jetta Frost, Margit Osterloh, and Antoinette Weibel, "Governing Knowledge Work: Transactional and Transformational Solutions," *Organizational Dynamics*, Vol. 39, No. 2, 2010, pp. 126–136.

[36] Frost, Osterloh, and Weibel, 2010.

[37] For a discussion of the relationship between structure and culture, and innovation, see Mark Ramsey and N. Barkhuizen, "Organisational Design Elements and Competencies for Optimising the Expertise of Knowledge Workers in a Shared Services Centre," *South African Journal of Human Resource Management*, Vol. 9, No. 1, 2011, pp. 158–172.

2. Optimizing the Alignment Between Work and STEM Professional Characteristics

Prior research has identified qualities of STEM work and the priorities of workers who are drawn to STEM fields that should be considered when developing and maintaining a STEM workforce: autonomy of STEM employees in selecting and managing their work, collaboration with specialists having complementary knowledge, focus on substantive work rather than management or administrative tasks, and flexible work arrangements (FWAs).[38] Compared with administrative or support workers, who can perform their work and are most productive in a structured, predictable (or routine) environment, STEM workers are more likely to thrive in a work environment that encourages creativity and innovation and allows them the mental space to experiment with new ideas or new ways of combining existing ones. Focused intellectual processes and the production of innovative ideas are at the core of STEM professionals' work activities. However, such intellectual and creative processes are unlikely to occur on-demand within a preset work schedule (Monday–Friday, 9:00 a.m.–5:00 p.m.), with many STEM researchers and scientists spending long hours in the laboratory and focused on their research and on conducting new experiments.

Knowing the characteristics, needs, and expectations of STEM professionals from their work environment helps organizations rethink and redesign the work setting, organizational culture, and climate that maximizes effort and fosters innovation. Building on research that shows that organizational "culture shapes the creation and adoption of new knowledge,"[39] it can be argued that aligning the work and the organization's culture and climate for a better fit with the known traits, attributes, and needs of the STEM professional enhances performance and overall organizational outcomes. Furthermore, acknowledging the individual and professional preferences in terms of culture and environment of women in STEM contributes to the efforts to design organizational policies that attract and retain a diverse pool of talent in STEM fields.[40]

---

[38] Charles D. Orth III, "The Optimum Climate for Industrial Research," in Norman Kaplan, ed., *Science and Society,* Chicago, Ill.: Rand-McNally, 1965, p. 141.

[39] Hayati Abdul Jalal, Paul Toulson, and David Tweed, "Exploring Employee Perceptions of the Relationships Among Knowledge Sharing Capability, Organizational Culture and Knowledge Sharing Success: Their Implications for HRM Practice," *Proceedings of the International Conference on Intellectual Capital, Knowledge Management and Organisational Learning,* January 2011, p. 640.

[40] While the authors acknowledge the underrepresentation of racial minorities across STEM occupations, in this report, we focus only on the case of women underrepresentation in STEM as it overlaps with their overall underrepresentation across DoD. That said, the retention of women in STEM occupations represents a dual challenge for the military. For those interested in an analysis of steps and initiatives taken to address the racial/ethnic diversity of DoD's STEM workforce, see Nelson Lim, Abigail Haddad, Dwayne M. Butler, and Kate Giglio, *First*

Organizational Culture and Climate

Organizational culture and climate are two distinct but closely related factors that influence the creativity and productivity of highly skilled workers such as STEM employees.[41] Charles Glisson contends that

> [c]urrent empirically based models of organizational innovation and effectiveness transcend the mechanistic models of a century ago and many emphasize that innovation and effectiveness are as much about creating the appropriate organizational social context as about implementing the latest technology. The idea that an organization's social context is associated with innovation and effectiveness is accepted by many organizational leaders and two distinct dimensions of social context—organizational culture and climate—are mentioned often as the key factors that determine an organization's performance in a wide range of areas.[42]

Hence, by internally nurturing an organizational culture and climate that take into account the particular work characteristics of STEM—autonomy, collaboration, focus on substantive work, and FWAs—an organization is more likely to promote creativity, innovation, and productivity across its STEM workforce.

Organizational culture and *climate* are concepts that have been debated extensively in the literature.[43] In the 1930s, Kurt Lewin referred to organizational climate as "the psychological impact of the work environment on employees' sense of well-being, motivation, behavior, and performance."[44] In the late 1970s and early 1980s,[45] "the shared behavioral norms, values, and expectations within an organization"[46] emerged to represent the organization's culture, which was a distinct concept from organizational climate.[47] On one hand, climate represents the workforce's shared perception of how the work environment psychologically impacts their well-

---

Steps Toward Improving DoD STEM Workforce Diversity, Santa Monica, Calif.: RAND Corporation, RR-329-OSD, 2013.

[41] Angelika Trübswetter, Karen Genz, Katharina Hochfeld, and Martina Schraudner, "Corporate Culture Matters—What Kinds of Workplaces Appeal to Highly Skilled Engineers?" *International Journal of Gender, Science and Technology*, Vol. 8, No. 1, 2016, pp. 46–66.

[42] Charles Glisson, "The Role of Organizational Culture and Climate in Innovation and Effectiveness," *Human Service Organizations: Management, Leadership and Governance*, Vol. 39, No. 4, 2015, pp. 245–250.

[43] Glisson, 2015.

[44] Kurt Lewin (1939) cited in Glisson, 2015.

[45] Daniel R. Denison, "What Is the Difference Between Organizational Culture and Organizational Climate? A Native's Point of View on a Decade of Paradigm Wars," *Academy of Management Review*, Vol. 21, No. 3, July 1996, p. 619.

[46] Handy (1976) and Pettigrew (1979) cited in Glisson, 2015.

[47] For additional definitions and discussions of the difference between organizational culture and climate, see Benjamin Schneider, Sarah K. Gunnarson, and Kathryn Niles-Jolly, "Creating the Climate and Culture of Success," *Organizational Dynamics*, Vol. 23, No. 1, Summer 1994, pp. 17–29; Denison, 1996; and Dov Zohar and David A. Hofmann, "Organizational Culture and Climate," in Steve W. J. Kozlowski, ed., *Oxford Handbook of Industrial and Organizational Psychology*, Vol. I, New York, N.Y.: Oxford University Press, 2012.

being and ability to perform in the workplace; on the other hand, culture focuses on how the organization's norms and expectations drive the manner in which the workforce engages with their everyday work.[48] Together, culture and climate address the organization's norms, how workers engage with the work, and the psychological impact of the work environment on workers.

Newton Margulies and Anthony Raia have identified various characteristics that STEM workers value in terms of organizational culture and climate, such as the organization's ability to provide "challenging and stimulating work assignments," "on-the-job colleague interaction . . . through both the formal task arrangements and informal discussions," "openness of communications," and "the extent to which flexible team effort is employed, and the autonomy of the individual scientist or engineer."[49]

Research conducted by Trübswetter et al. yielded similar findings, with highly skilled engineers describing their optimal organizational culture as "flexible, prioritizing work-life balance, employee-centered, empowering, and multi-cultural" and especially giving the employees "autonomy to determine when, where, and how they will work, including how they will distribute work among themselves."[50]

Organizational culture also influences knowledge sharing.[51] An organizational culture that encourages cooperation and informal meetings among employees facilitates knowledge sharing. Improvements in communication, cooperation, and the sharing of knowledge ultimately foster innovation, among other positive effects benefiting the organization, such as improved customer service and voluntarism.[52] Maxine Robertson and Jacky Swan found that STEM workers demanded high levels of autonomy and that an organizational culture that embraced "ambiguity"—defined as "a consensus that there would be no consensus"—provided the high-skilled workforce with the autonomy and flexibility they required to excel at their job.[53]

Hence, an organizational culture and climate that support STEM workers' autonomy, collaboration, focus on substantive work, and FWAs are more likely to attract and retain STEM employees and stimulate their innovative abilities and productivity. In the following subsections, we will turn to each of the four characteristics—autonomy, collaboration, focus on substantive work, and FWAs—and discuss how each of them contributes to stimulating innovation and productivity across the STEM workforce.

---

[48] Glisson, 2015; see also Trübswetter et al., 2016, p. 49.

[49] Newton Marguiles and Anthony P. Raia, "Scientists, Engineers, and Technological Obsolescence," *California Management Review*, Vol. 10, No. 2, December 1, 1967, pp. 44–46.

[50] Trübswetter et al., 2016, p. 52.

[51] Andrawina et al., 2008, and Kim and Lee, 2006, cited in Jalal, Toulson, and Tweed, 2011.

[52] Schneider, Gunnarson, and Niles-Jolly, 1994.

[53] Maxine Robertson and Jacky Swan, "'Control–What Control?' Culture and Ambiguity Within a Knowledge Intensive Firm," *Journal of Management Studies*, Vol. 40, No. 4, June 2003, p. 831.

Autonomy

Although individual differences are always a factor, knowledge workers in general and STEM professionals specifically have a higher need for autonomy.[54] As Drucker stated, "[b]ecause the 'players' in an information-based organization are specialists, they cannot be told how to do their work,"[55] and one of the more-notable characteristics of knowledge-based workers is their need for autonomy as related to performing the job itself.

The literature broadly describes different kinds of work autonomy and the relationship between these types of autonomy and employee satisfaction and performance. Lotte Bailyn draws a distinction between *strategic autonomy*, the ability to determine one's own research agenda, and *operational autonomy*, the ability to determine how one conducts his or her own research. Bailyn describes how the level of strategic and operational autonomy is best distributed among employees.[56] Positions with both high operational and high strategic autonomy are most frequently reserved for a few highly experienced employees who are expected to generate practical knowledge of benefit to the organization. Positions with higher strategic autonomy and lower operational autonomy are generally best for lab management or administrators. Most technical professionals that engage in lab work have greater operational autonomy with less strategic autonomy. Unsurprisingly, employees who are at the start of their careers or who are production oriented generally have low strategic and operational autonomy. Over the course of their careers, employees may move into positions with either higher strategic autonomy or higher operational autonomy, depending on their track.[57]

Similar to Bailyn, Donald Pelz and Frank Andrews focused on two aspects: (1) individual autonomy, which they described as one's ability to determine the goals and objectives of their technical work responsibilities (similar to strategic autonomy) and (2) coordination of situation (corresponding to some parts of operational autonomy), which they described as the amount of central management or coordination with a group or groups required to conduct work.[58] They examined the relationships between autonomy and coordination and their impact on performance and scientific contribution, noting that "from such a statistical analysis one cannot prove whether autonomy precedes and stimulates higher performance, or whether it is a reward given to those

---

[54] Robertson and Swan, 2003; Peter Ferdinand Drucker, "The New Society of Organizations," *Harvard Business Review*, September–October 1992.

[55] Drucker, 1988.

[56] Lotte Bailyn, "Autonomy in the Industrial R&D Lab," *Human Resource Management*, Vol. 24, No. 2, Summer 1985, pp. 129–146.

[57] Bailyn, 1985.

[58] Donald C. Pelz and Frank M. Andrews, "Autonomy, Coordination, and Stimulation, in Relation to Scientific Achievement," *Behavioral Science*, Vol. 11, No. 2, March 1966, pp. 89–97.

who have already achieved." However, they found more support for the first interpretation over the latter.[59] They added that

> autonomy appears to have been most beneficial to scientific contribution and organizational usefulness for persons in moderately tight [centrally managed] or mixed situations. As coordination diminished, autonomy may have been not only less helpful to achievement but may have actually hindered it.[60]

Pelz and Andrews posited that this decreased performance by highly autonomous individuals in loosely coordinated environments may be tied to isolation and exclusion of outside stimulation from colleagues, resulting in complacency and overly narrow specialization. They ultimately concluded that, in moderately coordinated situations, high autonomy was associated with high motivation and stimulation from interactions with peers. Furthermore, in these moderately coordinated situations, there was sufficient flexibility to allow motivation and peer interactions to support increased performance.[61]

Frank Harrison came to a similar conclusion: Performance improved in scientists who engaged in setting their own objectives and making decisions.[62] Similarly, George Miller concluded that STEM employees were more alienated from their work when working under a directive-supervision style (i.e., supervisor as decisionmaker and limited employee-supervisor interaction), whereas employees under participatory (i.e., joint decisionmaking and increased employee-supervisor interactions) and laissez-faire (i.e., employee as decisionmaker and limited employee-supervisor interaction) were less alienated.[63]

These examples from the literature show that autonomy can be highly motivating. Organizations might consider substituting autonomy for financial compensation as a motivator for knowledge workers because of the personal benefits that knowledge workers derive from making decisions about their own work.[64] Loss of autonomy, on the other hand, may be highly demotivating, as demonstrated in a case study of the effects of reduced autonomy on research and development (R&D) engineers at a global information technology (IT) company.[65] Pauline Gleadle, Damian Hodgson, and John Storey found that when an information technology

---

[59] Pelz and Andrews, 1966, p. 91.

[60] Pelz and Andrews, 1966, p. 92.

[61] Pelz and Andrews, 1966.

[62] Frank Harrison, "The Management of Scientists: Determinants of Perceived Role Performance," *Academy of Management Journal*, Vol. 17, No. 2, 1974, pp. 234–241.

[63] George A. Miller, "Professionals in Bureaucracy: Alienation Among Industrial Scientists and Engineers," *American Sociological Review*, Vol. 32, No. 5, October 1967, pp. 755–768.

[64] Alfonso Gambardella, Claudio Panico, and Giovanni Valentini, "Strategic Incentives to Human Capital," *Strategic Management Journal*, Vol. 36, No. 1, January 2015, pp. 37–52.

[65] Pauline Gleadle, Damian Hodgson, and John Storey, "'The Ground Beneath My Feet': Projects, Project Management and the Intensified Control of R&D Engineers," *New Technology, Work and Employment*, Vol. 27, No. 3, November 2012, pp. 163–177.

company had a change in management, the new management began exerting more centralized control over project- and portfolio-management objectives, frustrating the engineers working on the respective R&D projects. This eroded the previously supported culture that granted autonomy to the engineer experts who were working on various projects. In response, some engineers sought to join management to protect projects from being canceled, while others retreated within their projects or actively resisted the new management.[66] These findings are taken from a single case study, limiting their ability to be generalized across the STEM workforce overall, but they do offer an account of how engineers within one organization responded to their loss of autonomy.

In sum, autonomy comprises two components: (1) authority to select the focus of one's work and (2) authority to manage one's own work processes. Selecting the focus of one's work appears especially important to STEM workers and relates to higher performance. In fact, offering more autonomy in selecting work may motivate STEM professionals, while reducing autonomy may demotivate them. However, for military organizations such as the Air Force, where it might be more difficult for STEM employees to choose the focus of their work, allowing them to have a strong input into projects and autonomy in managing how they complete the work within the permitted security restrictions might represent a compensatory mechanism for the lack of autonomy in selecting work focus in most situations.

Collaboration and Work Design

STEM work relies on collaboration as a source of productivity and innovation.[67] As noted earlier, knowledge workers are likely to be productive and intellectually stimulated if they can collaborate with other specialists with complementary knowledge and skills. Collaboration may take the form of an individual consulting with a colleague, team-based projects involving multiple knowledge professionals, or even cross-pollination among multiple project teams. It might involve informal interactions, such as discussions over lunch, or formal activities such as meetings and peer review. Collaboration might occur entirely within an organization or might involve reaching out to external experts.

---

[66] Gleadle, Hodgson, and Storey, 2012.

[67] Giovanni Abramo, Ciriaco A. D'Angelo, and Flavia Di Costa, "Research Collaboration and Productivity: Is There Correlation?" *Higher Education*, Vol. 57, No. 2, 2009, pp. 155–171; Dries Faems, Bart Van Looy, and Koenraad Debackere, "Interorganizational Collaboration and Innovation: Toward a Portfolio Approach," *Journal of Product Innovation Management*, Vol. 22, No. 3, May 2005, pp. 238–250.

STEM Professionals Prioritize Collaboration

STEM workers, who are a subset of knowledge workers, rely heavily on learning from peers in the workplace.[68] STEM workers are usually frustrated when it is difficult for them to identify specific expert knowledge in the organization that would improve their job performance.[69] This is because they view personal relationships as important to transferring knowledge and consider on-the-job problem-solving and colleague interaction as important for their own professional growth.[70] Furthermore, work environments that are conducive to knowledge exchange increase morale, trust, and employee retention.[71]

Margulies and Raia reported that on-the-job colleague interaction is critical to research scientists and engineers. For professional growth, informal personal relationships and formal collaborative efforts were found to be second only to on-the-job problem-solving. Margulies and Raia concluded that "[t]he ease of building and maintaining informal relationships and networks of colleague interactions is seen as a significant characteristic of the organizational environment"[72] and is quintessential for the organization's success.

Marvel et al. have similar findings in their study of corporate entrepreneurship among scientists and engineers. They couple the need for collaboration and work design, finding that "[t]he job has to be structured right, which includes . . . working with other world-class technologists,"[73] while the daily interaction in the context of projects with "less-capable people is de-motivating."[74]

Cross-Team Collaboration

O'Leary, Mortensen, and Woolley examined the effects of working on multiple teams. Multiple team membership is common and an important factor in most STEM-oriented organizations.[75] Across a wide range of industries in both the United States and Europe, survey data report that about 65–95 percent of knowledge workers, including STEM workers, belong to

---

[68] Tam Yeuk-Mui May, Marek Korczynski, and Stephen J. Frenkel, "Organizational and Occupational Commitment: Knowledge Workers in Large Corporations," *Journal of Management Studies*, Vol. 39, No. 6, 2002, pp. 775–801.

[69] Ramsey and Barkhuizen, 2011.

[70] Margulies and Raia, 1967.

[71] Ramsey and Barkhuizen, 2011.

[72] Margulies and Raia, 1967, p. 44.

[73] Matthew R. Marvel, Abbie Griffin, John Hebda, and Bruce Vojak, "Examining the Technical Corporate Entrepreneurs' Motivation: Voices from the Field," *Entrepreneurship Theory and Practice*, Vol. 31, No. 5, September 2007, p. 762.

[74] Marvel et al., 2007, p. 762.

[75] Michael B. O'Leary, Mark Mortensen, and Anita Woolley, "Multiple Team Membership: A Theoretical Model of Its Effects on Productivity and Learning for Individuals and Teams," *Academy of Management Review*, Vol. 36, No. 3, 2011, pp. 461–478.

multiple project teams. O'Leary, Mortensen, and Woolley proposed a model to help scholars and managers understand which properties of job design are important in maximizing individual and organizational outcomes while mitigating any negative effects.[76]

Their findings point to two promising properties of job design. First, active coordination of schedules across teams appears to moderate the negative effects of multiple team membership on team productivity and learning. Nonoverlapping deadlines; contiguous blocks of time devoted to each project; and scheduling practices, such as fixed meeting times, are more beneficial. Second, they suggest clearly defining team roles, such as *core member* or *consultant*, so that expectations are set regarding each team member's priorities and meeting attendance.[77]

External Collaboration

Bruno Cassiman and Reinhilde Veugelers present a comprehensive overview of the literature on the complementarity of internal R&D and external collaboration, making a good case that successful innovation and competitive advantage result from this cross-fertilization.[78] Two components appear to be essential to fruitful outcomes from internal-external collaboration: The networks between the two must be well developed, and the internal R&D capability must be strong, because solid internal expertise is required to evaluate and apply external expertise to greatest effect.

External collaboration with other experts in the field often occurs in the context of conferences and annual meetings of professional organizations. According to Robert Hilborn, disciplinary societies and professional organizations "set the norms and expectations for professional work within the disciplines: what counts as research in the discipline, what are the standards for publication, and what professional behaviors are rewarded and recognized by others in the discipline?"[79] By participating in outside professional-development activities, such as professional meetings and conferences, STEM professionals also update their disciplinary knowledge and prevent the obsolescence of their skills and knowledge base.[80] Hence, conference and professional meeting participation allow STEM employees to stay current in their fields, and—for those involved in research—to remain visibly active in the research community and maintain their scientific credibility.

---

[76] O'Leary, Mortensen, and Woolley, 2011.

[77] O'Leary, Mortensen, and Woolley, 2011.

[78] Bruno Cassiman and Reinhilde Veugelers, "In Search of Complementarity in Innovation Strategy: Internal R&D and External Knowledge Acquisition," *Management Science*, Vol. 52, No. 1, January 2006, pp. 68–82.

[79] Robert C. Hilborn, "The Role of Scientific Societies in STEM Faculty Workshops Meeting Overview," *The Role of Scientific Societies in STEM Faculty Workshops: A Report of the May 3, 2012, Meeting*, Washington, D.C.: Council of Scientific Society Presidents, American Chemical Society, 2012, p. 13.

[80] Margulies and Raia, 1967, p. 43.

While attending such external professional-development events, STEM professionals also have the opportunity to exchange ideas and have their conceptual and empirical approaches validated or refuted by other experts in the field. Although such settings tend to be formal, they help foster both formal and informal interactions and exchanges and contribute to the expansion of the informal networks in which STEM professionals participate. Furthermore, for STEM researchers in particular, repeated formal and informal interactions with peers contribute to building mutual trust, often translating into interpersonal exchanges of resources across organizational boundaries, which are critical to fostering innovation.[81]

These observations and research findings point out that an organization's R&D environment must be intellectually rich, allowing for internal and external opportunities for robust collaboration.

Focus on Substantive Work

STEM workers are highly motivated by their daily work, and it is important for them to have responsibilities that they view as challenging and interesting.[82] In an early study on the topic, Charles Orth stated that "scientists and engineers cannot or will not . . . operate at the peak of their creative potential in an atmosphere that puts pressure on them to conform to organizational requirements which they do not understand or believe necessary,"[83] such as performing administrative tasks, which they are likely to perceive as intellectually dissatisfying routine work.

Similar findings were revealed in a recent study of STEM industry organizations. STEM workers most frequently cited a need for freedom in terms of time, and that they are highly demotivated when they have to spend their working hours on bureaucratic tasks.[84] Furthermore, James Kochanski and Gerald Ledford cite a "Rewards of Work" study, whose survey results demonstrated that 75 percent of scientific and technical talent reported that the quality of their work responsibilities directly influenced their retention with their current employer. In speaking about this population, they noted that "repetitive, narrow work with little individual discretion repels professionals."[85]

Along similar lines, Marvel et al. found that it is important for STEM workers to engage in intellectually challenging work and collaborate "on projects that have value to potential

---

[81] Isabelle Bouty, "Interpersonal and Interaction Influences on Informal Resource Exchanges Between R&D Researchers Across Organizational Boundaries," *Academy of Management Journal*, Vol. 43, No. 1, 2000, pp. 50–65.

[82] Marvel et al., 2007; James Kochanski and Gerald Ledford, "'How to Keep Me'—Retaining Technical Professionals," *Research-Technology Management*, Vol. 44, No. 3, May 2001, pp. 31–38.

[83] Orth, 1965, p. 141.

[84] Marvel et al., 2007.

[85] Kochanski and Ledford, 2001, p. 34.

customers." Moreover, in the same study, they discovered that the same workers found it demotivating "to work on mundane projects."[86]

Organizations can help improve STEM workers' motivation and interest in their work by offloading routine and administrative tasks onto lower-skilled specialists. Kochanski and Ledford note that, in an effort to improve retention of R&D professionals, some organizations make

> special efforts to reengineer their R&D jobs to eliminate, automate, or outsource routine tasks, and to make sure that staff have real decision-making rights and work in a collegial atmosphere. Another trend is to make sure that there are ways for staff to change assignments at least as easily as they can change employers, and to reduce the ability of a manager to hold staff in an assignment that they wish to leave.[87]

Kochanski and Ledford's findings encapsulate two of the important job characteristics to optimizing the STEM workforce: elimination of routine tasks and autonomy in choice of work.

Flexible Work Arrangements

Flexible work arrangements (FWAs) are most commonly defined as benefits offered by an employer that allow employees varying levels of control over the time during which and the place where work occurs.[88] Telecommuting, sometimes referred to as *flexplace*, is a type of FWA that permits employees to work from a location (such as home) other than the organizational facility.[89] Flextime gives employees some level of control over the hours during which they work during a day.[90] Some other types of FWAs include compressed workweeks, casual dress, mealtime flex, break arrangements, shift flexibility, seasonal scheduling, and job sharing.[91]

The large body of literature on FWAs indicates that they have considerable potential benefits. From the employee's perspective, FWAs allow for increased perception of autonomy and job

---

[86] Marvel et al., 2007, p. 762.

[87] Kochanski and Ledford, 2001, p. 37.

[88] Boris B. Baltes, Thomas E. Briggs, Joseph W. Huff, Julie A. Wright, and George A. Neuman, "Flexible and Compressed Workweek Schedules: A Meta-Analysis of Their Effects on Work-Related Criteria," *Journal of Applied Psychology*, Vol. 84, 1999, pp. 496–513; Ravi S. Gajendran and David A. Harrison, "The Good, the Bad, and the Unknown About Telecommuting: Meta-Analysis of Psychological Mediators and Individual Consequences," *Journal of Applied Psychology*, Vol. 92, No. 6, 2007, pp. 1524–1541; Alysa D. Lambert, Janet H. Marler, and Hal G. Gueutal, "Individual Differences: Factors Affecting Employee Utilization of Flexible Work Arrangements," *Journal of Vocational Behavior*, Vol. 73, No. 1, August 2008, pp. 107–117.

[89] Gajendran and Harrison, 2007.

[90] Rebecca J. Thompson, Stephanie C. Payne, and Aaron B. Taylor, "Applicant Attraction to Flexible Work Arrangements: Separating the Influence of Flextime and Flexplace," *Journal of Occupational and Organizational Psychology*, Vol. 88, No. 4, December 2015, pp. 726–749.

[91] Society for Human Resource Management, *SHRM Research: Flexible Work Arrangements*, Alexandria, Va., 2015.

satisfaction and decreased work-family conflict, job stress, and transportation costs.[92] From the employer's perspective, FWAs result in increased productivity, applicant attraction and retention, and decreased absenteeism and turnover intentions.[93]

However, regarding FWAs for STEM workers, the results are less clear-cut. In some ways, FWAs seem well suited for the STEM field because of its "need for a higher level of education and development" and "increased autonomy and responsibility" characteristic of typical STEM positions.[94] Gladys Hrobowski-Culbreath found that FWAs are good options when workers are able to schedule work that must be completed at the office on days when they do not telecommute and when their colleagues are satisfied with their use of FWAs.[95]

In other ways, however, FWAs are likely to not be a good fit for STEM work. According to Hrobowski-Culbreath, FWAs are ideal for project-based jobs where the main focus is for a worker to complete a task by a deadline with few other constraints.[96] While a lot of STEM work is project based,[97] the work is commonly a group effort requiring specific resources that are difficult to access, making both flextime and telecommuting difficult. Furthermore, for the DoD STEM workforce, the need to access classified information or perform lab work in classified locations makes the implementation of FWA challenging, with alternative work arrangements being needed to balance STEM workers' need for FWAs with office or lab presence.

FWAs were found to be beneficial when tasks are clearly defined with settable goals,[98] but work in the STEM field is rarely clearly defined from the beginning. Much of the issue with STEM workers taking advantage of FWAs comes from the common team-oriented aspect of STEM work. Coordinating with team members on nontelecommuting days can be difficult, and a

---

[92] Baltes et al., 1999; Gajendran and Harrison, 2007; Lambert, Marler, and Gueutal, 2008; Simone Kauffeld, Eva Jonas, and Dieter Frey, "Effects of a Flexible Work-Time Design on Employee- and Company-Related Aims," *European Journal of Work and Organizational Psychology*, Vol. 13, No. 1, 2004, pp. 79–100; Laurel A. McNall, Aline D. Masuda, and Jessica M. Nicklin, "Flexible Work Arrangements, Job Satisfaction, and Turnover Intentions: The Mediating Role of Work-to-Family Enrichment," *Journal of Psychology*, Vol. 144, No. 1, 2009, pp. 61–81; Lubica Bajzikova, Helena Sajgalikova, Emil Wojcak, and Michaela Polakova, "Are Flexible Work Arrangements Attractive Enough for Knowledge-Intensive Businesses?" *Procedia–Social and Behavioral Sciences*, Vol. 99, November 6, 2013, pp. 771–783; Kim and Gong, 2016.

[93] Baltes et al., 1999; Kauffeld, Jonas, and Frey, 2004; Gajendran and Harrison, 2007; Lambert, Marler, and Gueutal, 2008; McNall, Masuda, and Nicklin, 2009; Gladys Hrobowski-Culbreath, *Flexible Work Arrangements: An Evaluation of Job Satisfaction and Work-Life Balance*, dissertation, University of Missouri-Columbia, 2010, Columbia, Mo.: ProQuest Dissertations and Theses Database, 3423947, 2010; Thompson, Payne, and Taylor, 2015; Lisa M. Leslie, Colleen Flaherty Manchester, Tae-Youn Park, and Si Ahn Mehng, "Flexible Work Practices: A Source of Career Premiums or Penalties? "*Academy of Management Journal*, Vol. 55, No. 6, 2012, pp. 1407–1428.

[94] Bajzikova et al., 2013.

[95] Hrobowski-Culbreath, 2010.

[96] Hrobowski-Culbreath, 2010.

[97] Michael Bikard, Fiona E. Murray, and Joshua Gans, "Exploring Trade-offs in the Organization of Scientific Work: Collaboration and Scientific Reward," *Management Science*, Vol. 61, No. 7, July 2015, pp. 1473–1495; Frost, Osterloh, and Weibel, 2010.

[98] Hrobowski-Culbreath, 2010.

lack of face-to-face (FTF) interaction decreases the amount of information shared among coworkers.[99] When employees spend more than half of their time working from home, they see greater positive effects on work-family conflict, but they also often see a "deterioration of coworker relationships."[100]

Although telecommuting is a popular form of flexible work space, it is not the only approach. Creative approaches to in-person workspaces also can be designed around STEM workers' needs. The physical design of the places where knowledge work happens plays an important role in the productivity, innovation, and satisfaction of STEM workers, who need an environment that encourages collaboration and increases creativity translating to high-quality, innovative results.[101] To accomplish such an objective, workplaces need to be designed with this desired effect in mind.

There are a variety of office designs to choose from, and each can be customized to the needs of the organization and its STEM workers. Traditionally, workers have their own offices that are separated from others by walls and doors, referred to as *cell office spaces*.[102] This type of space can eliminate many of the distractions that can undermine creative knowledge work. In addition to office space, workplaces should offer spaces designed for different activities, such as collaboration space or production areas (multispace office designs).[103] There are also open-plan layouts, which are typically open areas with low (or no) walls and unassigned seating.[104] A less common but increasingly important office design is the urban hub, which provides a physical work location for employees who would typically telework. The urban hub should be conveniently located and offer tools and technical equipment necessary for workers to complete their jobs.[105] Urban hubs can be shared by multiple organizations to allow their workers to come together and share what might be expensive tools and technologies, as well as be in a modern workspace without having to travel to the typical office location.[106]

An important consideration when determining the ideal office design for a workforce is the amount of FTF interaction that occurs and whether this is something that would benefit the organization. FTF interaction has been found to result in greater information sharing among

[99] Kauffeld, Jonas, and Frey, 2004.

[100] Gajendran and Harrison, 2007.

[101] Bikard, Murray, and Gans, 2015.

[102] Roman Boutellier, Fredrik Ullman, Jurg Schreiber, and Reto Naef, "Impact of Office Layout on Communication in a Science-Driven Business," *R&D Management*, Vol. 38, No. 4, September 2008, pp. 372–391.

[103] Boutellier et al., 2008.

[104] Rianne Appel-Meulenbroek, "Knowledge Sharing Through Co-Presence: Added Value of Facilities," *Facilities*, Vol. 28, No. 3/4, 2010, pp. 189–205.

[105] Tammy Johns and Lynda Gratton, "The Third Wave of Virtual Work," *Harvard Business Review*, January–February 2013, pp. 66–73.

[106] Johns and Gratton, 2013.

knowledge workers, resulting in greater innovation and productivity.[107] A multispace office has nearly three times as much FTF interaction as cell office spaces.[108] Urban hubs also allow for greater FTF interaction, and they also facilitate external collaboration.[109]

Organizations are starting to realize the importance of investing in physical aspects of the work environment, with such companies as Google, Apple, and 3M investing substantial resources in creative work environments for their knowledge workers.[110] While organizations should consider their office design plan and potentially restructure the workspace, simpler physical elements can also help enhance creativity. These include but are not limited to less-crowded workspaces,[111] views of windows, and plants around the office, as well as the color, sound, and odor of the physical workspace.[112]

Women in STEM Fields

Autonomy, collaboration, focus on substantive work, and FWAs are also important for women in STEM. Alongside supporting these generic four characteristics of the STEM workforce, STEM organizational culture and climate also need to consider factors that have historically been demotivating for women STEM workers and incorporate changes in stereotypes associated with the STEM work environment and with women's abilities; increase the presence of women role models; and provide opportunities for professional growth, family-friendly policies, and a sexual harassment–free work environment.

Women continue to be underrepresented in some STEM occupations, such as engineering, computer, and physical sciences,[113] but also within DoD, where women have lower levels of

---

[107] Appel-Meulenbroek, 2010; Claudia E. Baumann, Frank Zoller, and Roman Boutellier, "Fostering Creativity and Innovation: Spheres of Interaction Influence Chance Encounters," in Carla Vivas and Fernando Lucas, eds., *Proceedings of the 7th European Conference on Innovation and Entrepreneurship*, Vol. I, Red Hook, N.Y.: Curran Associates, Inc., pp. 190–197.

[108] Boutellier et al., 2008.

[109] Johns and Gratton, 2013.

[110] Adam Brand, "Knowledge Management and Innovation at 3M," *Journal of Knowledge Management*, Vol. 2, No. 1, 1998, pp. 17–22; Roland Kuntze and Erika Matulich, "Google: Searching for Value," *Journal of Case Research in Business and Economics*, Vol. 2, May 2010, pp. 1–10; Stefan H. Thomke and Barbara Feinberg, "Design Thinking and Innovation at Apple," Harvard Business School Case 609-066, January 2009, pp. 1–14.

[111] John R. Aiello, Donna T. DeRisi, Yakov M. Epstein, and Robert A. Karlin, "Crowding and the Role of Interpersonal Distance Preference," *Sociometry*, Vol. 40, No. 3, 1977, pp. 271–282.

[112] Nancy J. Stone and Joanne M. Irvine, "Direct or Indirect Window Access, Task Type, and Performance," *Journal of Environmental Psychology*, Vol. 14, No. 1, March 1994, pp. 57–63; Seiji Shibata and Naoto Suzuki, "Effects of an Indoor Plant on Creative Task Performance and Mood," *Scandinavian Journal of Psychology*, Vol. 45, No. 5, 2004, pp. 373–381; Janetta Mitchell McCoy and Gary W. Evans, "The Potential Role of the Physical Environment in Fostering Creativity," *Creativity Research Journal*, Vol. 14, No. 3/4, 2010, pp. 409–426.

[113] Nikki Graf, Richard Fry, and Cary Funk, "7 Facts About the STEM Workforce," Pew Research Center, FactTank, January 9, 2018.

representation in the civilian workforce of each military service.[114] This overlap in underrepresentation for women who work in STEM occupations within DoD is worth a closer investigation to understand how the productivity of women in STEM occupations, especially of those who work in male-dominated environments such as the military, can be improved.

Research findings show "that culture and atmosphere in a workplace can substantially influence women's career decisions."[115] Moreover, STEM fields have their own "unique set of norms and values," and, in these fields, "individuals' likelihood of success . . . increases when they understand and adopt these norms and values."[116] As the STEM fields have been mostly controlled by a predominantly male workforce, the norms and structures in place are often exclusionary for women and translate into an "unwelcoming environment"[117] for many of them. Similar dynamics exist in military organizations such as the Air Force, where women are underrepresented in senior leadership roles.[118]

The shortage of women in STEM is mainly the result of challenges that STEM organizations face in attracting, recruiting, and retaining qualified women.[119] This phenomenon is mostly visible at the top of the hierarchy, where very few women occupy leadership positions,[120] either because they have dropped out along the way—the so-called leaky pipeline[121]—or they have been passed over for promotion in favor of male colleagues. The shortage of women in STEM fields is not exclusively a U.S. phenomenon; European and Asian countries experience similar trends of low participation of women in STEM.[122]

The large body of literature in industrial sociology that studies organizational culture and practices finds organizations to be usually "gendered and biased against women." These

---

[114] David Schulker and Miriam Matthews, *Women's Representation in the U.S. Department of Defense Workforce: Addressing the Influence of Veterans' Employment*, Santa Monica, Calif.: RAND Corporation, RR-2458-OSD, 2018, p. 2.

[115] Schramm and Kerst, 2009, and Singh et al., 2013, cited in Trübswetter et al., 2016, p. 50.

[116] McBride, 2018.

[117] Kimberly Griffin cited in McBride, 2018.

[118] Kirsten M. Keller, Kimberly Curry Hall, Miriam Matthews, Leslie Adrienne Payne, Lisa Saum-Manning, Douglas Yeung, David Schulker, Stefan Zavislan, and Nelson Lim, *Addressing Barriers to Female Officer Retention in the Air Force*, Santa Monica, Calif.: RAND Corporation, RR-2073-AF, 2018, p. 1.

[119] Amanda B. Diekman and Aimee L. Belanger, "New Routes to Recruiting and Retaining Women in STEM: Policy Implications of a Communal Goal Congruity Perspective," *Social Issues and Policy Review*, Vol. 9, No. 1, January 2015, p. 52; Jacob C. Blieckenstaff, "Women and Science Careers: Leaky Pipeline or Gender Filter?" *Gender and Education*, Vol. 17, No. 4, October 2005, p. 369.

[120] Diekman and Belanger, 2015, pp. 56, 75; Sandra A. Swanson, "Hidden in Plain Sight," *PM Network*, Vol. 28, No. 12, December 2014, pp. 42–49.

[121] For a more detailed discussion of the leaky pipeline metaphor, see Diekman and Belanger, 2015, and Blieckenstaff, 2005.

[122] Diekman and Belanger, 2015, p. 53; according to Swanson, 2014, p. 44, "women hold only 13 percent of all U.K. jobs" in STEM fields, while in South Korea they hold less than 15 percent of the jobs in engineering and technology.

organizational practices and biases are derived from the fact that most decisionmakers with impact on the organization's structure and culture are male.[123] Furthermore, recent literature on women in STEM outlines some of the key explanations regarding the shortage of women in STEM:

1. Differences in exposure to STEM fields and socialization during childhood and teenage years: Girls are less likely than boys to be exposed to and encouraged to develop STEM-related abilities in childhood and later in their teenage years.[124] The long-term impact is that fewer girls and young women decide to pursue STEM degrees and careers, narrowing the pipeline of available qualified women candidates.

2. Prevailing stereotypes: Fewer women pursue higher education degrees in STEM, and, among them, even fewer continue pursuing a STEM career because of prevailing stereotypes that STEM professions are not "feminine" but "masculine,"[125] and that STEM careers are less likely to allow women to build and nurture a family.[126] In addition, for the women who become part of the STEM workforce, negative gender stereotypes result in their being rated less competent when they engage intellectually with their male counterparts in the workplace, leading to the gradual erosion of their self-confidence and to professional disengagement, with some women eventually opting out of a STEM career path.[127]

3. Lack of successful women role models in STEM: The absence of role models defying existing negative stereotypes against women in STEM has an impact on the performance and retention of the women who have entered the field.[128] According to Drury, Siy, and Cheryan, the strong negative stereotypes women in STEM face result in high internal self-doubts about their ability to perform well in STEM fields. The erosion of self-confidence ultimately pushes many women out of STEM and leads those who remain on the job to underperform.[129] For these reasons, the presence of women role models in STEM not only "inoculates" other women STEM professionals "against the harmful effects of such negative stereotypes" but also prevents these women from underperforming and leaving STEM professions.[130]

---

[123] Acker, 1991, and Wetterer, 1995, cited in Trübswetter et al., 2016, pp. 49–50; Isis H. Settles, "Women in STEM: Challenges and Determinants of Success and Well-Being," American Psychological Association website, October 2014.

[124] Diekman and Belanger, 2015, p. 54.

[125] Diekman and Belanger, 2015, p. 55; McBride, 2018; Benjamin J. Drury, John Oliver Siy, and Sapna Cheryan, "When Do Female Role Models Benefit Women? The Importance of Differentiating Recruitment from Retention in STEM," *Psychological Inquiry*, Vol. 22, No. 4, 2011, pp. 265–269.

[126] Diekman and Belanger, 2015, p. 65; Drury, Siy, and Cheryan, 2011, p. 266; Erica S. Weisgram and Amanda Diekman, "Family-Friendly STEM: Perspectives on Recruiting and Retaining Women in STEM Fields," *International Journal of Gender, Science and Technology*, Vol. 8, No. 1, 2016, pp. 39–45.

[127] Diekman and Belanger, 2015, p. 68; Settles, 2014.

[128] Drury, Siy, and Cheryan, 2011.

[129] Drury, Siy, and Cheryan, 2011.

[130] Drury, Siy, and Cheryan, 2011, p. 265.

4. Sexual harassment of women in STEM: One often-cited reason for the leaky pipe among women in STEM is that "women are harassed out of science."[131] A June 2018 comprehensive report of the National Academies of Sciences, Engineering, and Medicine disclosed that close to half of "all women in science have experienced some form of sexual harassment."[132] As sexual harassment is more likely to happen in male-dominated work environments, women in STEM are more vulnerable to potential harassment than in fields with a more gender-balanced workforce.[133]

In light of these factors that—first and foremost—affect the well-being of women in STEM and ultimately have an impact on their retention, a work environment that is family friendly, where sexual harassment and negative stereotypes against women are absent and where other women can serve as active role models, is more likely to stimulate innovation and productivity among women in STEM. Commitment from management to fostering a work environment where women can thrive and the management's confidence in the abilities of the women hired and promoted within the organization are crucial in removing the existing negative stereotypes and in improving the self-confidence of women in the STEM workforce.[134]

Furthermore, research on workplace preferences of men and women has revealed that women place a higher value than men on work-life balance, FWAs, good hours, easy commutes, autonomy, interpersonal relationships, and professional growth.[135] Additionally, women in STEM generally value more "communal goals" (or orientation toward others),[136] which usually translate into a higher level of social interaction and collaboration across the organization. In this vein, increasing the perception and factual reality of family friendliness of the STEM field— more specifically the perception *and* reality that a STEM career will afford family goals—and "decreasing the baby-penalty that women with children pay,"[137] represent additional steps toward integrating, increasing the productivity, and fully developing the capabilities and talent of women in STEM.

In conclusion, to increase the productivity of women in STEM, the organizational culture and climate should include—alongside autonomy, collaboration, focus on substantive work, and FWA—the presence of STEM women role models and a stereotype- and sexual harassment–free work environment, as well as family-friendly policies and opportunities for professional growth.

---

[131] McBride, 2018.

[132] McBride, 2018.

[133] Settles, 2014.

[134] Swanson, 2014, p. 46–47.

[135] Trübswetter et al., 2016, p. 50; Swanson, 2014, p. 47.

[136] Diekman and Belanger, 2015, p. 61.

[137] Weisgram and Diekman, 2016, p. 42.

3. Human Capital Functions

Human capital functions are often portrayed by a life-cycle model, which takes place in seven major phases: workforce planning, talent acquisition, workforce development, performance management, rewards and recognition, career planning, and succession planning. Most of the key components of the human capital life cycle relate directly to the focus of this report. Development, rewards and recognition, career advancement, and performance management are all areas in which thoughtful and research-supported approaches can maximize the productivity of STEM professionals, especially if the approaches are well matched to the core four characteristics of STEM work. However, some elements of the human capital life cycle—workforce planning, talent acquisition, and succession planning—are outside the scope of this chapter, as they do not directly contribute to maximizing the productivity of the existing STEM workforce within an organization.

Development

Alongside compensation and benefits, STEM workers are motivated to perform well and remain with their employers by thoughtful human resource–management approaches with regard to career-development opportunities and education benefits.[138]

Professional growth is a powerful motivator for STEM workers. By providing opportunities for professional growth, an organization can create a more positive work environment and increase workers' self-motivation.[139] Research by Herman, Deal, and Ruderman found that federal employees are motivated by professional-development opportunities that can help lead to career advancement, and that organization-supported professional development is associated with higher work performance and productivity, as well as higher employee retention among federal professionals.[140] In this light, professional-development opportunities for federal STEM workers that allow them to acquire skills that contribute to career advancement are likely to improve their motivation and retention.

---

[138] David E. Frick, "Motivating the Knowledge Worker," Arlington, Va.: Defense Acquisition University, 2010, pp. 369–387.

[139] Margulies and Raia, 1967; Ramsey and Barkhuizen, 2011.

[140] Jeffrey L. Herman, Jennifer J. Deal, and Marian N. Ruderman, "Motivated by the Mission or by Their Careers?" *Public Manager*, Vol. 41, No. 2, Summer 2012, p. 43.

Although a lot of professional development takes place under the guise of on-the-job training, internal mentoring, coaching,[141] and through leadership feedback,[142] research conducted by Margulies and Raia finds that formal education courses are most effective at increasing STEM worker productivity if they relate to current organizational needs.[143]

As one approach to tailoring formal professional development to organizations' needs, some STEM organizations are leveraging professional science master's programs to build the skill sets they need in the STEM workforce. These advanced-degree programs combine STEM coursework with courses in project management, writing, and other skills that help students gain in-demand workplace and leadership experience for STEM careers. Initially created with funding from the Alfred P. Sloan Foundation in 2007, these degree programs usually take about two years to complete and are designed to be an alternative to academic STEM degrees.[144]

These degree programs are gaining traction and are offered at more than 100 colleges and universities across the country.[145] The curricula are frequently developed in consultation with STEM companies located near the schools to ensure that these programs continue to meet workplace needs. DoD could also work in conjunction with schools to develop coursework that specifically addresses DoD STEM needs to build a wider applicant pool, as well as to give current DoD STEM professionals the opportunity to return to school.[146] Not all professional-development opportunities need to be traditional classes, and it is important for organizations to emphasize professional growth through intra- and interorganization knowledge sharing and communication.[147]

Rewards and Recognition

Both extrinsic and intrinsic reward and recognition systems can significantly boost STEM workers' productivity. When properly implemented, rewards and recognition help strengthen

---

[141] For more general findings on this topic, please see the sections on "Building a Coaching Culture" and "Mentoring for Impact" in the "Global Leadership Forecast 2018: 25 Research Insights to Fuel Your People Strategy," by Development Dimensions International, Inc., The Conference Board Inc., EYGM Limited, 2018, pp. 36–39. It is unclear whether these findings pertain as well to the STEM workforce, with additional research being recommended.

[142] Frost, Osterloh, and Weibel, 2010.

[143] Margulies and Raia, 1967.

[144] Christopher J. Gearon, "Focus on Job Skills with a Professional Master's Degree," *U.S. News and World Report*, March 13, 2013.

[145] Gearon, 2013.

[146] National Academy of Engineering and National Research Council, 2012.

[147] Margulies and Raia, 1967; Ramsey and Barkhuizen, 2011.

STEM employees' efficiency and motivation, especially for those working in the federal government.[148]

Pay and Benefits

Examples of extrinsic rewards include pay, bonuses, and promotions. Rewards in the form of pay are a simple way to show organizational appreciation to employees, as well as boost employee motivation and commitment to the organization.[149] Similar to most knowledge workers, STEM workers are most motivated by individual rewards and are likely to leave a position if they are not rewarded for their performance.[150] It is also important to make it clear to STEM workers what they need to do to earn bonuses or pay increases.[151] This factor can actually be more important to STEM employees than the pay level itself.[152]

Government organizations can compete with STEM private-sector pay in two ways: (1) by minimizing the wage difference between public and private so that government pay stays above "the minimum level necessary to attract and retain the needed talent" and, "once that minimum level is achieved," (2) by emphasizing nonmonetary benefits and rewards, such as "quality of colleagues, quality and capability of facilities and quality of life."[153] In 2012 and 2017, the Congressional Budget Office reported that federal employees with professional and doctoral degrees earn 18 percent less than those employed in the private sector.[154] However, despite the difference between public and private wages, government research organizations, such as the National Institutes of Health, the National Institute of Standards and Technology, and the Naval Research Laboratory, have successfully retained STEM employees by providing—in addition to acceptable pay levels—indirect benefits such as excellent resources, facilities, and talent networks.[155]

---

[148] Herman, Deal, and Ruderman, 2012.

[149] Frost, Osterloh, and Weibel, 2010.

[150] Marvel et al., 2007.

[151] Kochanski and Ledford, 2001.

[152] Kochanski and Ledford, 2001.

[153] Coffey, 2008, p. 20.

[154] Congressional Budget Office, *Comparing the Compensation of Federal and Private-Sector Employees*, Washington, D.C., January 2012, p. ix, and Congressional Budget Office, *Comparing the Compensation of Federal and Private-Sector Employees, 2011–2015*, Washington, D.C., April 2017, p. 3.

[155] Coffey, 2008, pp. 19–20.

Internal Recognition

STEM employees are not primarily driven by money, unless the pay is obviously lower than for private-sector workers,[156] and organizations that rely too heavily on extrinsic rewards such as bonuses are unlikely to effectively motivate their STEM workforce, especially if the bonuses are perceived as a control mechanism of the employees.[157] STEM workers also desire feedback and recognition for their individual work and want to be rewarded for work that is unusual or achieves outcomes above and beyond their peers.[158] For example, Abbott Laboratories effectively takes advantage of this approach, and their internal reward system includes chairman's awards, president's awards, and patent or inventor awards to highlight outstanding achievements by employees.[159] Furthermore, STEM workers desire support and encouragement from their managers.[160] Management feedback can help STEM workers feel more competent and empowered in their work,[161] which is likely to translate into an increase in productivity.[162]

Internal recognition, if implemented incorrectly, can undermine productivity. STEM workers are demotivated when rewarded for normal, everyday work and when the rewards are distributed across the entire workforce rather than rewarding the most-accomplished employees.[163] Additionally, internal rewards—just like external ones—can be counterproductive if the workforce views them as controlling.[164]

Career Advancement

Across many organizations, promotion and transition into management are perceived as a sign of success. This stands true as well for STEM workers, who—according to Baylin—consider the lack of advancement, especially into management roles, as a sign of professional

---

[156] Regarding the role of financial compensation, we rely on Frederick Herzberg's two-factor theory, which states that while pay is not a strong motivator for workers, it can be a clear demotivating factor if it is not adequate for the work. See Frederick Herzberg, Bernard Mauser, and Barbara Bloch Snyderman, *The Motivation to Work*, New Brunswick, N.J., and London: Transaction Publishers, 2010.

[157] Frost, Osterloh, and Weibel, 2010.

[158] Marvel et al., 2007.

[159] Business Management Daily Editors, "Abbott Touts Innovation to Recruit, Retain Scientists," *Business Management Daily*, January 16, 2013.

[160] Marvel et al., 2007.

[161] Frost et al., 2010.

[162] Recent literature shows the positive relationship between employee and team empowerment and productivity; see Scott E. Seibert, Gang Wang, and Stephen H. Courtright, "Antecedents and Consequences of Psychological and Team Empowerment in Organizations: A Meta-Analytic Review," *Journal of Applied Psychology*, Vol. 96, No. 5, 2011.

[163] Marvel et al., 2007.

[164] Frost et al., 2010.

underachievement: "being an engineer after the age or 35 or 40 is considered failure."[165] However, not all engineers or STEM workers are alike, with some having a strong preference for the pursuit of a lifelong technical career, while others are administratively oriented. Across the two categories, Dirk Steiner and James Farr have identified three main groups of engineers with differing preferences and emphasis on promotion: (1) technically oriented individuals who rated promotion opportunities as being very important to them, (2) technically oriented individuals who rated promotion opportunities as being of low the importance to them, and (3) administratively career-oriented individuals for whom the importance of promotion was greater than for the technically oriented workforce.[166]

Steiner and Farr's study found that the second group—the low-promotion engineers—placed less value on being promoted into management roles than the two other groups. In light of their findings, when promotion and professional achievement across the STEM workforce of an organization are available mainly to those who embark on the managerial track with the technical track having fewer promotion options, there is a high risk that "management careers attract better performers away from the technical area where they are badly needed."[167]

To retain the technically oriented high performers in the areas of the organization in which they can contribute most, opportunities for career advancement and promotion should be equally available for both technical and managerial tracks, with the reward system for the former not lagging behind the latter in terms of status and financial compensation.

Performance Management

Performance management in a knowledge industry can be a complicated process because of the nature of knowledge work and the particular characteristics of STEM work.[168] As teamwork and collaboration within and across teams represent important aspects of the daily work life of STEM workers,[169] it is often difficult to disentangle and identify individual contributions of each team member. Individual assessments of performance are difficult because of the crossfertilization of ideas and the collective outcomes or outputs to which the entire team contributes. Moreover, for the technically focused but low promotion–inclined workers, performance-management practices that emphasize individual benefits (e.g., individual

---

[165] Baylin, 1980, cited in Dirk D. Steiner and James L. Farr, "Career Goals, Organizational Reward Systems and Technical Updating in Engineers," *Journal of Occupational Psychology*, Vol. 59, No. 1, March 1986, p. 14.

[166] Steiner and Farr, p. 17.

[167] Steiner and Farr, p. 21.

[168] This subsection mainly focuses on the knowledge worker more broadly because of the dearth of substantive research specific to human resources performance management and the STEM worker. However, as STEM workers are a subset of knowledge workers, the discussion is equally encompassing of STEM employees.

[169] We are aware that the extent of teamwork is likely to vary from job to job and from profession to profession. However, our underlying assumption is that most STEM work involves some degree of collaboration and teamwork.

promotions and bonuses[170]) are likely to prove counterproductive, while, for the remaining two categories (technically focused but promotion-inclined and administrative career–focused workers), performance systems that reward individual contribution are more likely to be motivating. [171]

Given the different individual inclinations of STEM workers toward pursuing a lifelong technically focused career or transitioning into management roles, as well as the different emphasis each individual places on promotion, the performance-management system should be tailored to take into account the individual's propensity, with the incentives mechanism designed to reward the areas in which the employees excel (technical or administrative).

---

[170] Frost, Osterloh, and Weibel, 2010.

[171] Steiner and Farr, 1986, p. 21.

4. The Role of Organizational Structure in Optimizing Performance of STEM Workers

As illustrated throughout this report, knowledge workers—and especially STEM professionals—look for and are more productive in certain types of working conditions. To this point, the discussion has focused on characteristics of the work and human resources system—qualities that could be addressed within an existing organizational structure. However, various organizational researchers have identified the growth of the knowledge industry as fundamentally different from prior industries and requiring fundamentally different organizational structures. In this section, we present the leading theories on how to organize knowledge-based organizations to maximize productivity. We begin by discussing the need to move from industrial- to knowledge-age structures and different approaches to management frameworks. We then explore the benefits of one particular approach that has achieved some currency in military and nonmilitary organizations: innovation cells. We conclude the section with a brief discussion of *hyperspecialization*, a phenomenon increasingly encountered within knowledge organizations and that involves narrowly specialized task forces or teams performing discrete tasks ultimately combined into a single knowledge product, such as software.

The Structure of Knowledge-Based Organizations

As noted earlier, industry has been moving away from the manufacturing to the knowledge-based business model. Drucker was among the first experts to explain why the knowledge-based organization must be structured differently from its predecessors:

> Information is data endowed with relevance and purpose. Converting data into information thus requires knowledge. And knowledge, by definition, is specialized. . . . The information-based organization requires far more specialists overall than the command-and-control companies are accustomed to.[172]

Drucker further posits that information-based organizations need to retain centralized functions such as legal and public relations,[173] but these service staffs will shrink drastically. The bulk of knowledge will be at the bottom of the organization, residing in the specialists who do the work and direct themselves.[174] Drucker also argues that knowledge-based organizations

---

[172] Drucker, 1988, pp. 46–47.

[173] Based on the definitions of *information-based organization* provided by Drucker in his 1988 article and on his discussion and interchangeable use of k*nowledge-based* and *information-based organization*, we also use interchangeably the terms *knowledge-based organization* and *information-based organization* in this report.

[174] Drucker, 1988, p. 47.

would be flatter, with fewer opportunities for employees to move into management. Therefore, the "pride and professionalism" of the specialists so critical to knowledge-based organizations would be maintained through the use of self-governing task forces.[175] Arising from this new structure driven by task forces with a small centralized staff, Drucker foresees a series of management problems arising, among them being the need to identify the management structure appropriate for these organizations and their respective business managers. The latter could be task force leaders or one of the two components of what Drucker terms "a two-headed monster" made up of a specialist structure and an administrative structure.[176]

As argued by Drucker, Lowell Bryan and Claudia Joyce also maintain that the traditional vertical organizational structure that emerged in the industrial age is neither efficient nor effective for the age of knowledge organizations. The two authors call for a full organizational structure redesign organized to maximize collaboration and knowledge sharing among professionals. This redesign can improve the efficiency of the work process, quality of the product, and satisfaction of the worker. Key elements of this redesign might include streamlined vertical structures, in which line managers focus on short-term earnings, knowledge professionals focus on long-term wealth development, organizational overlays are designed to improve collaboration, and performance-measurement approaches encourage professionals to self-direct their work toward goals rather than perform under close supervision.[177]

By moving away from the traditional hierarchical organization to a more decentralized and flat structure that connects, in a networklike fashion, autonomous task forces or units (e.g., innovation cells, see next section), a knowledge organization is more likely to increase the productivity of its STEM workforce. A networked structure within and across innovation cells is likely to not only stimulate innovation but also increase productivity by facilitating communication and collaboration, and—in this way—provide "timely access to knowledge and resources that are otherwise unavailable."[178]

Innovation Cells

An organization can create autonomous innovation cells to leverage its STEM workforce, increase productivity, and encourage innovation among its ranks. The terms *innovation cell* or *innovation corps* typically convey a separate, stand-alone unit that is structured differently, operates differently, and has different expectations for outcomes than its parent organization. Given the particular traits of knowledge organizations and knowledge workers, a simplified

---

[175] To Drucker, the term *task forces* means smaller self-governing units.

[176] Drucker, 1988, p. 51.

[177] Lowell Bryan and Claudia Joyce, "The 21st-Century Organization," *McKinsey Quarterly*, August 16, 2005.

[178] Walter W. Powell, Kenneth W. Koput, and Laurel Smith-Doerr, "Interorganizational Collaboration and the Locus of Innovation: Networks of Learning in Biotechnology," *Administrative Science Quarterly*, Vol. 41, No. 1, March 1996, p. 119.

vertical or even flat structure within the innovation cell itself and across the different cells is likely to have a multiplier effect.

Bryan and Joyce argue in favor of "the creation of enterprise-wide formal networks."[179] In their view, the creation of new products is the result of several multiyear projects that involve small groups of "full-time, focused professionals with the freedom to 'wander in the woods,' discovering new winning value propositions by trial and error and deductive tinkering." Bryan and Joyce also note the significance of a disciplined approach, observing that companies using this structure often allocate a fixed percentage of income to these long-term initiatives, assign top talent to work on these initiatives, and delegate a senior manager as sponsor.[180]

Innovation cells create a microlevel organizational culture and climate that supports autonomy, collaboration with like-minded experts or workers, focus on substantive work, and flexible work environment, which meets the core needs of STEM workers to excel and be productive. As the STEM workforce operates along different key characteristics from the rest of the workforce, especially those whose tasks are administrative in nature, the creation of innovation cells de facto separates the STEM workforce from others in a way that maximizes productivity and motivation of both groups of workers.[181]

DoD has already started to leverage innovation cells as a means to incorporate a start-up style flat and networked structure to fuel innovation and development. Within the Air Force, the 22nd Air Refueling Wing Plans and Programs Office has set up—among other innovation cells—XPX, an innovation team aimed to "produce homegrown, rapid solutions that will be implemented at the wing quickly and at low cost."[182] The U.S. Navy has created the Navy Innovation Cell to help target "industry's investments in technology and insert them into naval operations more quickly, tapping into innovation that can elude DoD."[183] Additionally, the Navy Innovation Cell is improving acquisition processes to better integrate emerging technologies into the Navy.[184]

In the context of innovation cells or narrowly specialized task forces found in modern-day knowledge organizations, a new phenomenon has surfaced: hyperspecialization.

---

[179] Bryan and Joyce, 2005; Lowell L. Bryan and Claudia I. Joyce, "Better Strategy Through Organizational Design," *McKinsey Quarterly*, No. 2, 2007.

[180] Bryan and Joyce, 2005.

[181] The explanation is that most lab researchers and scientists have irregular work hours, working late into the night to finish a task or experiment and, at times, having a later start of their workday. Witnessing the late arrival to work of STEM research workers often demotivates administrative staff, who are typically unaware of the long hours researchers spend in the lab after the end of the official work day at 5:00 p.m. The resulting frictions often have a demotivating effect on both workforces, with their separation through the creation of innovation cells reducing the potential for friction, loss of motivation, and lessened productivity.

[182] Erin McClellan, "Wing Stands Up Innovation Cell," McConnell Air Force Base, Kan.: U.S. Air Force Expeditionary Center, January 11, 2018.

[183] Amber Corrin, "Navy's Innovation Cell Fast-Tracks New Technologies," C4ISRNET.com, March 27, 2015.

[184] Corrin, 2015.

Hyperspecialization is perceived as increasing productivity and product quality, although there are some concerns that, in the long run, it might end up stifling innovation. Given that hyperspecialization is closely associated with the flat, networked nature of modern knowledge organizations, in the following subsection, we discuss briefly what hyperspecialization is, its impact on STEM productivity and innovation, and why it is likely that its benefits outweigh the risks.

Hyperspecialization

In the 18th century, Adam Smith first advanced the concept of division of labor,[185] which has since governed the way in which work is structured across organizations. However, in recent decades, advances in technology and communications and the advent of knowledge work have transformed division of labor, leading to hyperspecialization. In the context of the traditional division of labor, for example, factory workers in assembly lines carried out specialized tasks, ultimately assembling a physical product such as a car. In the context of hyperspecialization, 21st-century knowledge workers connect to their employers through technology and carry out discrete tasks remotely, which are then combined into a knowledge product such as software.

For example, software can be developed through hyperspecialization, beginning with the design phase.[186] A company might hold a competition to acquire the best new software product ideas. Having selected an idea (and rewarded the winner), the company could solicit proposals for the design specification and then for the design architecture. The company could separately solicit coders to produce each component in the software, as well as an expert to integrate the pieces. Finally, the company could hold multiple competitions for experts to identify and resolve bugs in each section of the software.

There are various models for hyperspecialization.[187] Organizations might outsource their specialized tasks to specific suppliers (under a contractual agreement or a similar legal arrangement), tap into a community of freelance workers (often through an intermediary organization), or develop their own in-house team of specialized knowledge workers. They might seek specialized support for low-level repetitive tasks (such as telemarketing) or advanced, expert knowledge tasks, such as solving a conceptual problem.

Knowledge workers may be particularly drawn to aspects of hyperspecialized work. They may have more control over their own work and their work-life balance, taking tasks that most interest them and fit with their schedules. They can work anywhere around the world rather than

---

[185] Adam Smith, *The Wealth of Nations*, New York, N.Y.: Modern Library, [1776] 2000.

[186] Thomas W. Malone, Robert Laubacher, and Tammy Johns, "The Big Idea: The Age of Hyperspecialization," *Harvard Business Review*, July–August 2011.

[187] Malone, Laubacher, and Johns, 2011; Xenios Thrasyvoulou, "Embracing Freelancers and the Age of Hyperspecialization," Relevance.com, June 16, 2015.

be restricted by geography and may be compensated at higher rates than other employees or, in low-wage areas, their neighbors.

Organizations may also benefit from this type of work setup, especially in terms of quality, speed, and cost. Quality is likely to be high: Knowledge workers who are competing for pieces of work, based largely on their prior efforts, are incentivized to maximize their performance. Furthermore, by specializing in a specific knowledge task, they provide expertise that a generalist does not always offer, and organizations that can access a large pool of individuals with this expertise may surface unique ways of thinking about a problem. Hyperspecialization can also be faster than having individuals carry out all elements of a project, especially if the subtasks can be carried out concurrently rather than sequentially. Finally, costs may be lower because the organization is not paying for the expert to learn something new—only for the production of the knowledge product—and is not paying for benefits or unproductive time. And hyperspecialization models that involve competition require payment only for successful products, reducing waste.

Establishing a hyperspecialization approach to work does require some additional activities and may entail some risk. Organizations need to break down larger tasks into discrete subtasks, being careful to group subtasks that have interdependencies. They must recruit their specialized workers, whether in-house or external, and establish vehicles for the business relationship (e.g., contracts, incentives). They need to establish a strategy for reintegrating the subproducts and have quality-control mechanisms; they may choose to source these two tasks to specialists as well.

On a broader note, hyperspecialization is a relatively new approach—and one with international reach. The lack of consistent regulations to govern the work across topics and countries may lead to abuses by both employers and specialists. Concerns have also been raised that hyperspecialization may inhibit innovation. However, despite the additional activities involved and potential risk, hyperspecialization is being explored widely and may change the approach to knowledge work.

5. Conclusions and Recommendations

In our review of the existing scholarly and professional literature on knowledge organizations and STEM workers' productivity and innovation, we found that STEM workers are more productive and innovative when organizational culture and climate promote four key characteristics: autonomy, collaboration, focus on substantive work, and FWAs. The review of the literature also revealed that women in STEM need role models provided by other women who have succeeded in their respective fields, family-friendly policies, opportunities for professional growth, and a work environment free of negative stereotypes and of sexual harassment.

In terms of human-capital functions, extending the professional-development opportunities of the civilian STEM workforce and tailoring the rewards, recognition, and performance-management systems to match individual inclinations and interest in technical versus management tracks are likely to optimize the performance of the existing STEM workforce within the Air Force. As for the organizational structure most conducive to stimulating productivity and innovation across the STEM workforce, our review of the literature indicated that the structure most likely to provide the STEM employees with a culture and climate fostering autonomy, collaboration, focus on substantive work, and flexibility was a network of autonomous cells or task forces. While there is evidence that these aspects are likely to improve the productivity of STEM workers in general, we recommend that the Air Force conduct its own independent study to determine which of the factors and in what combinations are likely to have the highest impact on the productivity of civilian STEM workers.

Aligning Work and STEM Professionals' Characteristics

By promoting an organizational culture and climate that take into account the particular characteristics of STEM work such as autonomy, collaboration, focus on substantive work, and FWAs, the Air Force is more likely to promote creativity, innovation, and productivity across its civilian STEM workforce. In addition, to fully benefit from the skills and capability that the women in the STEM workforce contribute, the Air Force should consider addressing the factors that demotivate women in STEM occupations and take into account the women-specific drivers that are likely to optimize their productivity.

Furthermore, the Air Force should consider putting in place a mechanism that strikes a balance between providing STEM workers with FWAs as needed, complemented by meeting in-office requirements (such as the manipulation of classified information), and access to the organization's facilities and technologies. Because STEM workers in the Air Force and in other DoD components are government employees who generally have to deal with more administrative tasks than their counterparts in the private sector, the Air Force might want to unburden them from bureaucratic minutiae and allow them to focus on research or on performing the substantive STEM-related work for which they were trained and which keeps them intellectually engaged.

Human Capital Functions

Regarding human capital functions, we recommend that the Air Force expand the professional-development opportunities offered to civilian STEM employees. In addition to increasing employee effectiveness, expanding education programs to the civilian STEM workforce could help boost retention. DoD already effectively uses postdoctoral fellowships to attract STEM professionals who have just received their terminal degree, and these fellowships frequently lead to participants continuing employment with DoD. Similarly, the Air Force could use the existing Air Force Science and Technology Fellowship Program as a cost-efficient and fast way to draw newly graduated talent into the organization. Likewise, the Air Force could also host fellowships that are funded by other federal organizations, such as the U.S. Department of Homeland Security.[188]

Concerning the rewards and recognition system, because STEM workers are motivated and improve their performance when their individual contributions to the organization are recognized, the Air Force should consider implementing and extending a contribution-based system to those parts of the organization where such a system is not already present. Furthermore, promoting a culture in which STEM employees receive periodic feedback from their supervisors is likely to increase the STEM workers' productivity.

Finally, the Air Force might consider bringing the compensation of its STEM workforce in line as much as possible with private-sector compensation, while allowing for autonomy and flexibility, as well as performance-management and career-advancement paths that take into account each individual's interests in promotion and in the pursuit of the available career tracks.

Organizational Structure Optimizing the Performance of STEM Professionals

In light of our findings related to the organizational structure best suited to increasing the productivity of STEM employees, we recommend that the Air Force set in place separate, simplified—even flat—structures that facilitate collaboration and knowledge sharing across the STEM workforce.

Setting up autonomous cells or task forces (similar to the 22nd Air Refueling Wing) that interact with one another in a networked, nonhierarchical way is likely to provide the STEM

---

[188] National Academy of Engineering and National Research Council, 2012.

workers who are involved in research with autonomy and control over their own work priorities and research agenda. A high level of autonomy for researchers and other STEM employees increases their productivity and is likely to translate into new, innovative ideas. A networked, nonhierarchical structure linking various innovation cells to one another across the organization would provide STEM employees with access to other experts and knowledge workers with whom they can collaborate and exchange ideas—crucial interactions that improve the effectiveness of STEM workers. A flat and interconnected organizational structure touches simultaneously on two of the four core factors involved in improving the productivity of the STEM workforce: autonomy and collaboration with specialists having complementary knowledge.

Challenges to Implementation

We acknowledge that the implementation of these recommendations is likely to encounter various barriers associated with the hierarchical and bureaucratic nature of the Air Force. As a quintessential command-and-control organization, the Air Force's culture and climate will need to adjust to support an organizational work environment in which STEM employees experience autonomy; focus on substantive, nonadministrative work; and are permitted FWAs. The need to safeguard critical information for national security purposes adds an additional challenge to the implementation of FWAs, as does allowing STEM employees autonomy to manage their own work processes and to choose the focus of their work. However, the Air Force's strong focus on promoting teamwork is likely to facilitate collaboration across its workforce, including its STEM professionals.

Given these inherent organizational culture and climate barriers, setting up separate, autonomous innovation cells in which the rules of the game can be rewritten to match the work characteristics of STEM professionals and to allow for more autonomy, flexibility, and focus on nonadministrative tasks is likely to optimize their performance.

Finally, we acknowledge that, in addition to the organizational barriers reflected in its culture and climate, the Air Force also may face statutory barriers in implementing some of the recommendations associated with human-capital functions. Such statutory barriers are likely to be related to limitations in existing government policies concerning the professional development, compensation, rewards, recognition, performance management, and career advancement of the civilian workforce. As the current statutory provisions associated with civilian compensation and professional development are mandated by Congress, a major overhaul of the existing statutes may be necessary to facilitate the implementation of needed changes in human-capital functions to optimize the productivity of civilian STEM professionals.

Selected Bibliography

Abramo, Giovanni, Ciriaco A. D'Angelo, and Flavia Di Costa, "Research Collaboration and Productivity: Is There Correlation?" *Higher Education*, Vol. 57, No. 2, 2009, pp. 155–171.

Aiello, John R., Donna T. DeRisi, Yakov M. Epstein, and Robert A. Karlin, "Crowding and the Role of Interpersonal Distance Preference," *Sociometry*, Vol. 40, No. 3, 1977, pp. 271–282.

Appel-Meulenbroek, Rianne, "Knowledge Sharing Through Co-Presence: Added Value of Facilities," *Facilities*, Vol. 28, No. 3/4, 2010, pp. 189–205.

Bailyn, Lotte, "Autonomy in the Industrial R&D Lab," *Human Resource Management*, Vol. 24, No. 2, Summer 1985, pp. 129–146.

Bajzikova, Lubica, Helena Sajgalikova, Emil Wojcak, and Michaela Polakova, "Are Flexible Work Arrangements Attractive Enough for Knowledge-Intensive Businesses?" *Procedia— Social and Behavioral Sciences*, Vol. 99, November 6, 2013, pp. 771–783.

Baltes, Boris B., Thomas E. Briggs., Joseph W. Huff, Julie A. Wright, and George A. Neuman, "Flexible and Compressed Workweek Schedules: A Meta-Analysis of Their Effects on Work-Related Criteria," *Journal of Applied Psychology*, Vol. 84, 1999, pp. 496–513.

Baumann, Claudia E., Frank Zoller, and Roman Boutellier, "Fostering Creativity and Innovation: Spheres of Interaction Influence Chance Encounters," in Carla Vivas and Fernando Lucas, eds., *Proceedings of the 7th European Conference on Innovation and Entrepreneurship*, Vol. I, Red Hook, N.Y.: Curran Associates, Inc., 2012, pp. 190–197.

Bikard, Michael, Fiona E. Murray, and Joshua Gans, "Exploring Trade-offs in the Organization of Scientific Work: Collaboration and Scientific Reward," *Management Science*, Vol. 61, No. 7, July 2015, pp. 1473–1495.

Blieckenstaff, Jacob C., "Women and Science Careers: Leaky Pipeline or Gender Filter?" *Gender and Education*, Vol. 17, No. 4, October 2005, pp. 369–386.

Boutellier, Roman, Fredrik Ullman, Jurg Schreiber, and Reto Naef, "Impact of Office Layout on Communication in a Science-Driven Business," *R&D Management*, Vol. 38, No. 4, September 2008, pp. 372–391.

Bouty, Isabelle, "Interpersonal and Interaction Influences on Informal Resource Exchanges Between R&D Researchers Across Organizational Boundaries," *Academy of Management Journal*, Vol. 43, No. 1, 2000, pp. 50–65.

Brand, Adam, "Knowledge Management and Innovation at 3M," *Journal of Knowledge Management*, Vol. 2, No. 1, 1998, pp. 17–22.

Bryan, Lowell, and Claudia Joyce, "The 21st-Century Organization," *McKinsey Quarterly*, August 16, 2005.

Bryan, Lowell L., and Claudia I. Joyce, "Better Strategy Through Organizational Design," *McKinsey Quarterly*, No. 2, 2007. As of November 15, 2019: http://www.michaelsamonas.gr/images/Mixalhs/resources/Better_Strategy.pdf

Business Management Daily Editors, "Abbott Touts Innovation to Recruit, Retain Scientists," *Business Management Daily*, January 16, 2013. As of November 15, 2019: https://www.businessmanagementdaily.com/34137/abbott-touts-innovation-to-recruit-retain-scientists/

Cassiman, Bruno, and Reinhilde Veugelers, "In Search of Complementarity in Innovation Strategy: Internal R&D and External Knowledge Acquisition," *Management Science*, Vol. 52, No. 1, January 2006, pp. 68–82.

Coffey, Timothy, "Building the S&E Workforce for 2040: Challenges Facing the Department of Defense," Washington, D.C.: Center for Technology and National Security Policy, National Defense University, July 2008. As of December 12, 2019: https://www.hsdl.org/?view&did=27341

Congressional Budget Office, *Comparing the Compensation of Federal and Private-Sector Employees*, Washington, D.C., January 2012. As of December 12, 2019: https://www.cbo.gov/sites/default/files/112th-congress-2011-2012/reports/01-30-fedpay0.pdf

Congressional Budget Office, *Comparing the Compensation of Federal and Private-Sector Employees, 2011–2015*, Washington, D.C., April 2017. As of December 12, 2019: https://www.cbo.gov/system/files/115th-congress-2017-2018/reports/52637-federalprivatepay.pdf

Corrin, Amber, "Navy's Innovation Cell Fast-Tracks New Technologies," C4ISRNET.com, March 27, 2015. As of November 13, 2019: https://www.c4isrnet.com/it-networks/2015/03/27/navy-s-innovation-cell-fast-tracks-new-technologies/

Crawford, Kate, Helen M. Hasan, Leoni Warne, and Henry Linger, "From Traditional Knowledge Management in Hierarchical Organizations to a Network Centric Paradigm for a Changing World," *Emergence: Complexity and Organization*, Vol. 11, No. 1, 2009, pp. 1–18.

Denison, Daniel R., "What Is the Difference Between Organizational Culture and Organizational Climate? A Native's Point of View on a Decade of Paradigm Wars," *Academy of Management Review*, Vol. 21, No. 3, July 1996, pp. 619–654.

Development Dimensions International, Inc., the Conference Board Inc., and EYGM Limited, "Global Leadership Forecast 2018: 25 Research Insights to Fuel Your People Strategy," 2018.

Diekman, Amanda B., and Aimee L. Belanger, "New Routes to Recruiting and Retaining Women in STEM: Policy Implications of a Communal Goal Congruity Perspective," *Social Issues and Policy Review*, Vol. 9, No. 1, January 2015, pp. 52–88.

Dovey, Ken, and Bryan Fenech, "The Role of Enterprise Logic in the Failure of Organizations to Learn and Transform: A Case from the Financial Services Industry," *Management Learning*, Vol. 38, No. 5, 2007, pp. 573–590.

Drucker, Peter F., "The Coming of the New Organization," *Harvard Business Review*, January 1988. As of November 13, 2019:
https://hbr.org/1988/01/the-coming-of-the-new-organization

Drucker, Peter F., "The New Society of Organizations," *Harvard Business Review*, September–October 1992. As of November 13, 2019:
https://hbr.org/1992/09/the-new-society-of-organizations

Drury, Benjamin J., John Oliver Siy, and Sapna Cheryan, "When Do Female Role Models Benefit Women? The Importance of Differentiating Recruitment from Retention in STEM," *Psychological Inquiry*, Vol. 22, No. 4, 2011, pp. 265–269.

Dul, Jan, and Canan Ceylan, "The Impact of a Creativity-Supporting Work Environment on a Firm's Product Innovation Performance," *Journal of Product Innovation Management*, Vol. 31, No. 6, November 2014, pp. 1254–1267.

Emerson, "Emerson Survey: 2 in 5 Americans Believe the STEM Worker Shortage Is at Crisis Level," August 21, 2018. As of November 19, 2018:
https://www.emerson.com/en-us/news/corporate/2018-stem-survey

Faems, Dries, Bart Van Looy, and Koenraad Debackere, "Interorganizational Collaboration and Innovation: Toward a Portfolio Approach," *Journal of Product Innovation Management*, Vol. 22, No. 3, May 2005, pp. 238–250.

Feller, Rich, "Advancing the STEM Workforce Through STEM-Centric Career Development," *Technology and Engineering Teacher*, Vol. 71, No. 1, September 2011, pp. 6–12.

Frick, David E., "Motivating the Knowledge Worker," Arlington, Va.: Defense Acquisition University, 2010, pp. 369–387.

Frost, Jetta, Margit Osterloh, and Antoinette Weibel, "Governing Knowledge Work: Transactional and Transformational Solutions," *Organizational Dynamics*, Vol. 39, No. 2, 2010, pp. 126–136.

Gajendran, Ravi S., and David A. Harrison, "The Good, the Bad, and the Unknown About Telecommuting: Meta-Analysis of Psychological Mediators and Individual Consequences," *Journal of Applied Psychology*, Vol. 92, No. 6, 2007, pp. 1524–1541.

Gambardella, Alfonso, Claudio Panico, and Giovanni Valentini, "Strategic Incentives to Human Capital," *Strategic Management Journal*, Vol. 36, No. 1, January 2015, pp. 37–52.

Gearon, Christopher J., "Focus on Job Skills with a Professional Master's Degree," *U.S. News and World Report*, March 13, 2013. As of November 13, 2019: https://www.usnews.com/education/best-graduate-schools/articles/2013/03/13/focus-on-job-skills-with-a-professional-masters-degree

Gleadle, Pauline, Damian Hodgson, and John Storey, "'The Ground Beneath My Feet': Projects, Project Management and the Intensified Control of R&D Engineers," *New Technology, Work and Employment*, Vol. 27, No. 3, November 2012, pp. 163–177.

Glisson, Charles, "The Role of Organizational Culture and Climate in Innovation and Effectiveness," *Human Service Organizations: Management, Leadership and Governance*, Vol. 39, No. 4, 2015, pp. 245–250.

Graf, Nikki, Richard Fry, and Cary Funk, "7 Facts About the STEM Workforce," Pew Research Center, FactTank, January 9, 2018. As of November 13, 2019: https://www.pewresearch.org/fact-tank/2018/01/09/7-facts-about-the-stem-workforce/

Harrington, Lisa M., Lindsay Daugherty, S. Craig Moore, and Tara L. Terry, *Air Force–Wide Needs for Science, Technology, Engineering, and Mathematics (STEM) Academic Degrees*, Santa Monica, Calif.: RAND Corporation, RR-659-AF, 2014. As of November 13, 2019: https://www.rand.org/pubs/research_reports/RR659.html

Harris, Richard G., "The Knowledge-Based Economy: Intellectual Origins and New Economic Perspectives," *International Journal of Management Reviews*, Vol. 3, No. 1, March 2001, pp. 21–40.

Harrison, Frank, "The Management of Scientists: Determinants of Perceived Role Performance," *Academy of Management Journal*, Vol. 17, No. 2, 1974, pp. 234–241.

Hausknecht, John P., Julianne Rodda, and Michael J. Howard, "Targeted Employee Retention, Performance-Based and Job-Related Differences in Reported Reasons for Staying," *Human Resources Management*, Vol. 48, No. 2, March–April 2009, pp. 269–288.

Herman, Jeffrey L., Jennifer J. Deal, and Marian N. Ruderman, "Motivated by the Mission or by Their Careers?" *Public Manager*, Vol. 41, No. 2, Summer 2012.

Herzberg, Frederick, Bernard Mauser, and Barbara Bloch Snyderman, *The Motivation to Work*, New Brunswick, N.J., and London: Transaction Publishers, 2010.

Hilborn, Robert C., "The Role of Scientific Societies in STEM Faculty Workshops: Meeting Overview," *The Role of Scientific Societies in STEM Faculty Workshops: A Report of the May 3, 2012, Meeting*, Washington, D.C.: Council of Scientific Society Presidents, American Chemical Society, 2012.

Hrobowski-Culbreath, Gladys, *Flexible Work Arrangements: An Evaluation of Job Satisfaction and Work-Life Balance*, dissertation, University of Missouri–Columbia, 2010, Columbia, Mo.: ProQuest Dissertations and Theses Database, 3423947, 2010.

Jalal, Hayati Abdul, Paul Toulson, and David Tweed, "Exploring Employee Perceptions of the Relationships Among Knowledge Sharing Capability, Organizational Culture and Knowledge Sharing Success: Their Implications for HRM Practice," *Proceedings of the International Conference on Intellectual Capital, Knowledge Management and Organisational Learning*, January 2011, pp. 639–646.

Johns, Tammy, and Lynda Gratton, "The Third Wave of Virtual Work," *Harvard Business Review*, January–February 2013, pp. 66–73. As of November 13, 2019: https://hbr.org/2013/01/the-third-wave-of-virtual-work

Kaplan, Fred, "The Pentagon's Innovation Experiment," *MIT Technology Review*, Vol. 120, No. 1, December 19, 2016, pp. 64–71. As of November 13, 2019: https://www.technologyreview.com/s/603084/the-pentagons-innovation-experiment/

Kastelle, Tim, "Hierarchy Is Overrated," *Harvard Business Review*, November 20, 2013. As of November 13, 2019: https://hbr.org/2013/11/hierarchy-is-overrated

Kauffeld, Simone, Eva Jonas, and Dieter Frey, "Effects of a Flexible Work-Time Design on Employee- and Company-Related Aims," *European Journal of Work and Organizational Psychology*, Vol. 13, No. 1, 2004, pp. 79–100.

Keller, Kirsten M., Kimberly Curry Hall, Miriam Matthews, Leslie Adrienne Payne, Lisa Saum-Manning, Douglas Yeung, David Schulker, Stefan Zavislan, and Nelson Lim, *Addressing Barriers to Female Officer Retention in the Air Force*, Santa Monica, Calif.: RAND Corporation, RR-2073-AF, 2018. As of November 13, 2019: https://www.rand.org/pubs/research_reports/RR2073.html

Keum, Dongil D., and Kelly E. See, "The Influence of Hierarchy on Idea Generation and Selection in the Innovation Process," *Organization Science*, Vol. 28, No. 4, July–August 2017, pp. 653–669.

Kim, Hyondong, and Yaping Gong, "Effects of Work-Family and Family-Work Conflicts on Flexible Work Arrangements Demand: A Gender Role Perspective," *International Journal of Human Resource Management*, Vol. 28, No. 20, November 2017, pp. 1–21.

Kochanski, James, and Gerald Ledford, "'How to Keep Me'—Retaining Technical Professionals," *Research-Technology Management*, Vol. 44, No. 3, May 2001, pp. 31–38.

Kuntze, Roland, and Erika Matulich, "Google: Searching for Value," *Journal of Case Research in Business and Economics*, Vol. 2, May 2010, pp. 1–10.

Lambert, Alysa D., Janet H. Marler, and Hal G. Gueutal, "Individual Differences: Factors Affecting Employee Utilization of Flexible Work Arrangements," *Journal of Vocational Behavior*, Vol. 73, No. 1, August 2008, pp. 107–117.

Leslie, Lisa M., Colleen Flaherty Manchester, Tae-Youn Park, and Si Ahn Mehng, "Flexible Work Practices: A Source of Career Premiums or Penalties?" *Academy of Management Journal*, Vol. 55, No. 6, 2012, pp. 1407–1428.

Lim, Nelson, Abigail Haddad, Dwayne M. Butler, and Kate Giglio, *First Steps Toward Improving DoD STEM Workforce Diversity*, Santa Monica, Calif.: RAND Corporation, RR-329-OSD, 2013. As of November 13, 2019:
https://www.rand.org/pubs/research_reports/RR329.html

Lipnack, Jessica, and Jeffrey Stamps, *The Age of the Network: Organizing Principals for the 21st Century*, Essex Junction, Vt.: Oliver Wight Publications, Inc., 1994.

Malone, Thomas W., Robert Laubacher, and Tammy Johns, "The Big Idea: The Age of Hyperspecialization," *Harvard Business Review*, July–August 2011. As of November 13, 2019:
https://hbr.org/2011/07/the-big-idea-the-age-of-hyperspecialization

Margulies, Newton, and Anthony P. Raia, "Scientists, Engineers, and Technological Obsolescence," *California Management Review*, Vol. 10, No. 2, December 1, 1967, pp. 43–48.

Marvel, Matthew R., Abbie Griffin, John Hebda, and Bruce Vojak, "Examining the Technical Corporate Entrepreneurs' Motivation: Voices from the Field," *Entrepreneurship Theory and Practice*, Vol. 31, No. 5, September 2007, pp. 753–768.

May, Tam Yeuk-Mui, Marek Korczynski, and Stephen J. Frenkel, "Organizational and Occupational Commitment: Knowledge Workers in Large Corporations," *Journal of Management Studies*, Vol. 39, No. 6, 2002, pp. 775–801.

McBride, Lisa, "Changing the Culture for Women and Underrepresented Groups in STEM+M," *Insights into Diversity*, August 22, 2018. As of November 13, 2019:
https://www.insightintodiversity.com/changing-the-culture-for-women-and-underrepresented-groups-in-stemm/

McClellan, Erin, "Wing Stands Up Innovation Cell," McConnell Air Force Base, Kan.: U.S. Air Force Expeditionary Center, January 11, 2018. As of November 13, 2019:

https://www.expeditionarycenter.af.mil/News/Article-Display/Article/1417920/wing-stands-up-innovation-cell/

McNall, Laurel A., Aline D. Masuda, and Jessica M. Nicklin, "Flexible Work Arrangements, Job Satisfaction, and Turnover Intentions: The Mediating Role of Work-to-Family Enrichment," *Journal of Psychology*, Vol. 144, No. 1, 2009, pp. 61–81.

Miller, George A., "Professionals in Bureaucracy: Alienation Among Industrial Scientists and Engineers," *American Sociological Review*, Vol. 32, No. 5, October 1967, pp. 755–768.

Mitchell McCoy, Janetta, and Gary W. Evans, "The Potential Role of the Physical Environment in Fostering Creativity," *Creativity Research Journal*, Vol. 14, No. 3/4, 2010, pp. 409–426.

National Academy of Engineering and National Research Council, *Assuring the U.S. Department of Defense a Strong Science, Technology, Engineering, and Mathematics (STEM) Workforce*, Washington, D.C.: National Academies Press, 2012.

Nonaka, Ikujiro, and Hirotaka Takeuchi, *The Knowledge-Creating Company: How Japanese Companies Create the Dynamics of Innovation*, New York and Oxford, UK: Oxford University Press, 1995.

Office of the Under Secretary of Defense for Acquisition and Sustainment, and Office of the Deputy Assistant Secretary of Defense for Manufacturing and Industrial Base Policy, *Report to Congress: Fiscal Year 2017 Annual Industrial Capabilities*, Washington, D.C., March 2018.

O'Leary, Michael Boyer, Mark Mortensen, and Anita Williams Woolley, "Multiple Team Membership: A Theoretical Model of Its Effects on Productivity and Learning for Individuals and Teams," *Academy of Management Review*, Vol. 36, No. 3, 2011, pp. 461–478.

Orth, Charles D. III, "The Optimum Climate for Industrial Research," in Norman Kaplan, ed., *Science and Society*, Chicago, Ill.: Rand-McNally, 1965, p. 141.

Pelz, Donald C., and Frank M. Andrews, "Autonomy, Coordination, and Stimulation, in Relation to Scientific Achievement," *Behavioral Science*, Vol. 11, No. 2, March 1966, pp. 89–97.

Powell, Walter W., Kenneth W. Koput, and Laurel Smith-Doerr, "Interorganizational Collaboration and the Locus of Innovation: Networks of Learning in Biotechnology," *Administrative Science Quarterly*, Vol. 41, No. 1, March 1996, pp. 116–145.

Ramsey, Mark, and Nicolene Barkhuizen, "Organisational Design Elements and Competencies for Optimising the Expertise of Knowledge Workers in a Shared Services Centre," *South African Journal of Human Resource Management*, Vol. 9, No. 1, 2011, pp. 158–172.

Robertson, Maxine, and Jacky Swan, "'Control—What Control?' Culture and Ambiguity Within a Knowledge Intensive Firm," *Journal of Management Studies*, Vol. 40, No. 4, June 2003, pp. 831–858.

Schneider, Benjamin, Sarah K. Gunnarson, and Kathryn Niles-Jolly, "Creating the Climate and Culture of Success," *Organizational Dynamics*, Vol. 23, No. 1, Summer 1994, pp. 17–29.

Schulker, David, and Miriam Matthews, *Women's Representation in the U.S. Department of Defense Workforce: Addressing the Influence of Veterans' Employment*, Santa Monica, Calif.: RAND Corporation, RR-2458-OSD, 2018. As of November 13, 2019: https://www.rand.org/pubs/research_reports/RR2458.html

Settles, Isis H., "Women in STEM: Challenges and Determinants of Success and Well-Being," American Psychological Association Science Brief, October 2014. As of January 31, 2020: https://www.apa.org/science/about/psa/2014/10/women-stem

Shibata, Seiji, and Naoto Suzuki, "Effects of an Indoor Plant on Creative Task Performance and Mood," *Scandinavian Journal of Psychology*, Vol. 45, No. 5, 2004, pp. 373–381.

Smith, Adam, *The Wealth of Nations*, New York: Modern Library, [1776] 2000.

Society for Human Resource Management, *SHRM Research: Flexible Work Arrangements*, Alexandria, Va., 2015. As of December 11, 2019: https://www.shrm.org/hr-today/trends-and-forecasting/special-reports-and-expert-views/Documents/Flexible%20Work%20Arrangements.pdf

Steiner, Dirk D., and James L. Farr, "Career Goals, Organizational Reward Systems and Technical Updating in Engineers," *Journal of Occupational Psychology*, Vol. 59, No. 1, March 1986, pp. 13–24.

Stone, Nancy J., and Joanne M. Irvine, "Direct or Indirect Window Access, Task Type, and Performance," *Journal of Environmental Psychology*, Vol. 14, No. 1, March 1994, pp. 57–63.

Sutherland, Margie, and Wilhelm Jordaan, "Factors Affecting the Retention of Knowledge Workers," *South African Journal of Human Resource Management*, Vol. 2, No. 2, 2004, pp. 55–64.

Swanson, Sandra A., "Hidden in Plain Sight," *PM Network*, Vol. 28, No. 12, December 2014, pp. 42–49.

Thomas, Will, "FY19 Budget Request: Defense S&T Stable as DOD Focuses on Technology Transition," American Institute of Physics, No. 20, February 23, 2018. As of November 19, 2018: https://www.aip.org/fyi/2018/fy19-budget-request-defense-st-stable-dod-focuses-technology-transition

Thomke, Stefan H., and Barbara Feinberg, "Design Thinking and Innovation at Apple," *Harvard Business School Case* 609-066, January 2009, pp. 1–14.

Thompson, Rebecca J., Stephanie C. Payne, and Aaron B. Taylor, "Applicant Attraction to Flexible Work Arrangements: Separating the Influence of Flextime and Flexplace," *Journal of Occupational and Organizational Psychology*, Vol. 88, No. 4, December 2015, pp. 726–749.

Thrasyvoulou, Xenios, "Embracing Freelancers and the Age of Hyperspecialization," Relevance.com, June 16, 2015. As of November 13, 2019:
https://www.relevance.com/embracing-freelancers-and-the-age-of-hyperspecialization/

Trübswetter, Angelika, Karen Genz, Katharina Hochfeld, and Martina Schraudner, "Corporate Culture Matters—What Kinds of Workplaces Appeal to Highly Skilled Engineers?" *International Journal of Gender, Science and Technology*, Vol. 8, No. 1, 2016, pp. 46–66.

U.S. Congress Joint Economic Committee, *STEM Education: Preparing for the Jobs of the Future*, a report by the Joint Economic Committee Chairman's Staff, Senator Bob Casey, Chairman, Washington, D.C., April 2012.

U.S. Department of Labor, U.S. Bureau of Labor Statistics, "Employment Projections: Employment in STEM Occupations 2018–2028," last updated September 4, 2019. As of December 6, 2019:
https://www.bls.gov/emp/tables/stem-employment.htm

Vilorio, Dennis, "STEM 101: Intro to Tomorrow's Jobs," *Occupational Outlook Quarterly*, Spring 2014, pp. 2–12.

Weisgram, Erica S., and Amanda Diekman, "Family-Friendly STEM: Perspectives on Recruiting and Retaining Women in STEM Fields," *International Journal of Gender, Science and Technology*, Vol. 8, No. 1, 2016, pp. 39–45.

Wenger, Etienne, "Communities of Practice and Social Learning Systems: The Career of a Concept," in Chris Blackmore, ed., *Social Learning Systems and Communities of Practice*, London: Springer-Verlag, 2010, pp. 179–198.

The White House, *National Security Strategy of the United States*, Washington, D.C., May 2010.

The White House, *National Security Strategy of the United States of America*, Washington, D.C., December 2017. As of November 20, 2018:
https://www.whitehouse.gov/wp-content/uploads/2017/12/NSS-Final-12-18-2017-0905.pdf

Xue, Yi, and Richard C. Larson, "STEM Crisis or STEM Surplus: Yes and Yes," *Monthly Labor Review*, Vol. 138, May 2015. As of February 4, 2020:
https://www.bls.gov/opub/mlr/2015/article/stem-crisis-or-stem-surplus-yes-and-yes.htm

Zohar, Dov, and David A. Hofmann, "Organizational Culture and Climate," in Steve W. J. Kozlowski, ed., *The Oxford Handbook of Industrial and Organizational Psychology*, Vol. 1, New York: Oxford University Press, 2012.